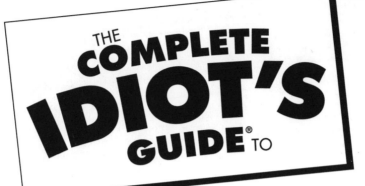

# Pre-Algebra

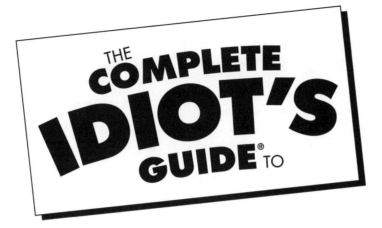

THE **COMPLETE IDIOT'S GUIDE**® TO

# Pre-Algebra

*by Amy F. Szczepanski, Ph.D.,
and Andrew P. Kositsky*

ALPHA

A member of Penguin Group (USA) Inc.

## ALPHA BOOKS

Published by the Penguin Group

Penguin Group (USA) Inc., 375 Hudson Street, New York, New York 10014, U.S.A.

Penguin Group (Canada), 10 Alcorn Avenue, Toronto, Ontario, Canada M4V 3B2 (a division of Pearson Penguin Canada Inc.)

Penguin Books Ltd, 80 Strand, London WC2R 0RL, England

Penguin Ireland, 25 St Stephen's Green, Dublin 2, Ireland (a division of Penguin Books Ltd)

Penguin Group (Australia), 250 Camberwell Road, Camberwell, Victoria 3124, Australia (a division of Pearson Australia Group Pty Ltd)

Penguin Books India Pvt Ltd, 11 Community Centre, Panchsheel Park, New Delhi—110 017, India

Penguin Group (NZ), cnr Airborne and Rosedale Roads, Albany, Auckland 1310, New Zealand (a division of Pearson New Zealand Ltd)

Penguin Books (South Africa) (Pty) Ltd, 24 Sturdee Avenue, Rosebank, Johannesburg 2196, South Africa

Penguin Books Ltd, Registered Offices: 80 Strand, London WC2R 0RL, England

International Standard Book Number: 978-1-59257-772-9
Library of Congress Catalog Card Number: 2008920828

10 09 08     8 7 6 5 4 3 2 1

Interpretation of the printing code: The rightmost number of the first series of numbers is the year of the book's printing; the rightmost number of the second series of numbers is the number of the book's printing. For example, a printing code of 08-1 shows that the first printing occurred in 2008.

*Printed in the United States of America*

**Note:** This publication contains the opinions and ideas of its author. It is intended to provide helpful and informative material on the subject matter covered. It is sold with the understanding that the author and publisher are not engaged in rendering professional services in the book. If the reader requires personal assistance or advice, a competent professional should be consulted.

The author and publisher specifically disclaim any responsibility for any liability, loss, or risk, personal or otherwise, which is incurred as a consequence, directly or indirectly, of the use and application of any of the contents of this book.

Most Alpha books are available at special quantity discounts for bulk purchases for sales promotions, premiums, fundraising, or educational use. Special books, or book excerpts, can also be created to fit specific needs.

For details, write: Special Markets, Alpha Books, 375 Hudson Street, New York, NY 10014.

**Publisher:** *Marie Butler-Knight*
**Senior Managing Editor:** *Billy Fields*
**Editorial Director/Acquiring Editor:** *Mike Sanders*
**Development Editor:** *Ginny Munroe*
**Production Editor:** *Megan Douglass*
**Copy Editor:** *Drew Patty*

**Cartoonist:** *Richard King*
**Cover Designer:** *Bill Thomas*
**Book Designer:** *Trina Wurst*
**Indexer:** *Angie Bess*
**Layout:** *Chad Dressler*
**Proofreader:** *John Etchison*

# Contents at a Glance

**Part 1:**     **The Building Blocks of Math**     1

   1   The Whole Story     3
*We start you off with a review of some old concepts and give you an overview of some of the more important new concepts you'll need for this book.*

   2   Filling In the Number Line     19
*Integers just aren't enough to run an international Wildebeest Cartel, and you'll need plenty of cents of dollars to go from the counting numbers to a full number line.*

   3   Properties of Number Systems     31
*Unfortunately, math is not a democracy and in order to participate effectively there are a number of laws and customs you can't break.*

**Part 2:**     **Let's All Remain Rational**     45

   4   Getting to Know Fractions     47
*Do you like long division? If not, fractions are for you!*

   5   Tips on Percentages     63
*Although your first introduction to percentage tipping may not be as exciting as your first experience with cow tipping, we think you'll enjoy learning about percents nonetheless, if only for their use in everyday life.*

   6   Ratio and Proportion     79
*Whether you're interested in model planes, mini-footballs, cooking, or science, rations and proportions will help you understand these subjects and more.*

**Part 3:**     **Radical and Exponential Power**     91

   7   Integral Exponents     93
*This is true power that no one can take away from you. Exponents grow faster than almost anything else, and with the information here you can start to wield their strength.*

   8   Roots and Other Exponents     107
*Whether square or potato-based, roots are yummy and delicious indeed. With these, you can cut exponents down to size and gain control of their mighty power.*

9   Decimals, Addition, and Subtraction Meet Exponents      125
*These are some of the applications and real-life uses of
exponents. You'll get plenty of introduction to the practical
results as well as the theory of topics like scientific notation.*

**Part 4:   Self-Expression Through Algebra      139**

10   It's Just an Expression      141
*In this chapter you can see what those silly letters are all
about in mathematics.*

11   Introduction to Equations      157
*Equality is one of the golden rules of mathematics. Solving
equations is the route to fame, fortune, money, and
algebra.*

12   Introduction to Algebra      175
*Algebra is the next step in mathematics, and after finish-
ing this chapter you'll be ready for the equation-based part
of it.*

**Part 5:   Ancient Foundations: Geometry      189**

13   Shapes on a Plane      191
*We take you back to kindergarten with connect-the-dots.
Only this time, you're in charge of drawing the dots as well.*

14   Angles in the Outfield      211
*Whenever you hear a bell ring, an angle gets formed.
You'll be introduced to some very special angles and how to
measure them.*

15   Poly(gon) Gets a Cracker      229
*Though you've seen squares, triangles, hexagons, and octa-
gons in everyday life, you'll get to know about polygons and
their evil pirate plots like never before.*

16   Area, Volume, and Surface Area      247
*All three of the major topics in this chapter deal with how
much stuff you have either in or on something. Turns out
you've got a lot of stuff indeed!*

17   Proportions      265
*Last I checked, buying a model plane didn't require a pilot's
license or an airport hanger.*

**Part 6:**    **Data Analysis**      **281**

18   Collecting Data      283
*Whether you're taking attendance in class or counting the number of headaches you've gotten from math, you're dealing with data and measurements.*

19   Probably Probability      303
*There's a very good chance that we talk about probability in this chapter. How good? Better than Tom Cruise's hair.*

**Appendixes**

A   Glossary      313

B   Solutions to Practical Practice Problems      321

Index      359

# Contents

**Part 1:  The Building Blocks of Math                                1**

**1  The Whole Story                                                  3**

Natural Numbers ..........................................................................4

Factors and Multiples ..................................................................5

*Prime Factorizations* ..............................................................7

*Greatest Common Factor* ........................................................9

*Least Common Multiple* .......................................................10

Integers ......................................................................................12

The Number Line and Absolute Value ......................................13

What's Your Sign? ....................................................................14

*Adding and Subtracting* ........................................................14

*Multiplying and Dividing* .....................................................16

**2  Filling In the Number Line                                     19**

Getting Down with Decimals ....................................................20

*Decimals Make the World Go Round* .....................................20

*Keeping It Real* ....................................................................21

Starting Rationally ....................................................................22

*Introduction to Fractions* ....................................................23

*Anatomy of a Fraction* .........................................................24

Lining Up on the Number Line ................................................25

*Numerical Comparison* .........................................................25

*Decimals on the Number Line* ...............................................26

*Fractions on the Number Line* ..............................................27

Rational vs. Irrational Numbers ...............................................28

**3  Properties of Number Systems                                  31**

Order of Operations ..................................................................32

*Grouping Symbols* .................................................................32

*The Next Order of Business* ...................................................33

Commutative Law ......................................................................34

Associative Law .........................................................................36

Distributive Law ........................................................................38

*Standard Problems* ...............................................................38

*Distributing a Minus Sign* ....................................................39

Identity Element ........................................................................41

Inverse Operations ....................................................................43

**Part 2:  Let's All Remain Rational                                45**

**4  Getting to Know Fractions                                     47**

Talking Rationally About Fractions.................................48
Secret Agent Fraction ....................................................50
  *Equivalent Fractions*.................................................*50*
  *Reducing to Simplest Terms* ....................................*52*
Fractions and Decimals...................................................53
  *Terminating Decimals* ..............................................*54*
  *Repeating Decimals*..................................................*55*
Multiplying and Dividing Fractions.................................56
Adding and Subtracting Fractions....................................58
  *Fractions with Like Denominators* .............................*58*
  *Fractions with Unlike Denominators*...........................*59*

**5  Tips on Percentages                                           63**

What Are Percents? .......................................................64
  *Percents and Decimals*..............................................*64*
  *Percents and Fractions* .............................................*66*
Calculating with Percents ...............................................68
  *Finding Percentages* .................................................*68*
  *Percent Change*........................................................*70*
Financial Examples ........................................................71
  *Interest*....................................................................*71*
  *Sales Tax*..................................................................*73*
  *Tipping*.....................................................................*73*
  *Commissions*.............................................................*74*
  *Profit* .......................................................................*75*
Percent Error................................................................76

**6  Ratio and Proportion                                          79**

Keeping Everything in Proportion ...................................80
Cooking the Night Away ................................................82
  *Mixed Drinks*............................................................*82*
  *Me Want Cookie!*.......................................................*83*
Unit Multipliers............................................................84
  *Show Me the Money: Converting Currency*.....................*85*
  *Furlongs per Fortnight: Converting Units*.......................*87*

**Part 3:    Radical and Exponential Power**    **91**

**7    Integral Exponents**    **93**

The Power of Exponents ................................................ 94

*Repeated Multiplication* ............................................ 94

*Writing Integers in Exponential Form* ...................... 96

*Order of Operations Reloaded* ................................... 97

Multiplying with Same Base ........................................ 98

*Positive Bases* ............................................................ 98

*Negative Bases* .......................................................... 99

*Can You Make an Exception?* ................................. 101

Negative Exponents .................................................. 101

Dividing with Same Base .......................................... 103

More Confusing Matters with Bases .......................... 104

**8    Roots and Other Exponents**    **107**

Defining Square Roots .............................................. 108

Calculating Square Roots .......................................... 109

*Perfect Squares* ....................................................... 110

*Imperfect Squares* ................................................... 111

*Dealing with Radicals* ............................................ 114

Don't Be Square: Other Roots .................................. 114

Rationalizing the Denominator .................................. 116

*A Single Square Root* .............................................. 116

*Two Square Roots* ................................................... 117

Rational Exponents ................................................... 118

*Simply Squares, Cubes, and Such* ........................... 119

*Powers and Roots* .................................................... 119

*Negative Rational Exponents* .................................. 120

*Let Our Powers Combine* ........................................ 121

Using Exponents to Simplify Fractions ..................... 122

*Answering in Fraction Form* .................................... 122

*Using Positive and Negative Exponents* ................... 123

**9    Decimals, Addition, and Subtraction Meet Exponents**    **125**

Base Ten ................................................................... 126

Scientific Notation .................................................... 127

*Significant Figures* .................................................. 128

*Very Big Numbers* ................................................... 128

*Very Small Numbers* ............................................... 129

Financial Math ............................................................. 131
    *Investment*.............................................................. *131*
    *Endowments*........................................................... *132*
    *Mortgages*.............................................................. *133*
Mathemagical Connections ............................................. 134
    *Areas of Squares*...................................................... *135*
    *Solving Problems with Primes*...................................... *136*
    *A Magical Mystery Tour of Numbers*............................ *136*

**Part 4:   Self-Expression Through Algebra                     139**

**10   It's Just an Expression                                  141**

Where Have the Numbers Gone? ..................................... 142
    *Variable Names* ...................................................... *142*
    *Naming Traditions* .................................................. *143*
Mathematical Expressions .............................................. 144
    *Operations*............................................................. *145*
    *Polynomials* ........................................................... *146*
Combining Like Terms ................................................. 147
Simplifying Products .................................................... 149
    *Multiplying Monomials*............................................. *149*
    *Using the Distributive Law* ....................................... *150*
    *FOILed Again* ....................................................... *150*
Plugging In: Numbers Return ........................................ 153
Found in Translation ................................................... 154

**11   Introduction to Equations                                157**

What Are Equations?.................................................... 158
Equations with One Type of Operation ............................ 160
    *Addition and Subtraction*........................................... *161*
    *Multiplication and Division* ....................................... *162*
Multi-Step Equations.................................................... 163
    *More Than One Operation* ........................................ *163*
    *Combining Like Terms* ............................................. *164*
Equations with More Than One Variable .......................... 165
Equations with More Than One Solution .......................... 167
    *Polynomials Reloaded* ............................................... *168*
    *Multiple Variables with Various Powers* ........................ *169*
    *Beyond Polynomials*................................................. *171*

Checking Your Work ................................................................ 172

   *Single Variable* .............................................................. 172

   *Multiple Variables* ........................................................ 173

**12 Introduction to Algebra** **175**

Simplifying with Fractions ..................................................... 176

   *Simplifying with Common Factors* ............................... 176

   *Positive or Negative Exponents?* .................................... 177

Cross-Multiplying .................................................................. 178

Working Together .................................................................. 180

Planes, Trains, and Automobiles ............................................ 182

**Part 5: Ancient Foundations: Geometry** **189**

**13 Shapes on a Plane** **191**

Plotting Points ...................................................................... 192

   *Remembering the Number Line—Cartesian Coordinates* .......... 192

Graphing Equations ............................................................... 197

Measuring Distance ............................................................... 201

Four Types of Rigid Symmetry .............................................. 203

   *Slip Sliding Away* .......................................................... 204

   *Turn, Turn, Turn* .......................................................... 206

   *Into the Looking Glass* ................................................... 207

   *Footprints in the Sand* ................................................... 209

**14 Angles in the Outfield** **211**

Measuring Angles ................................................................. 212

   *Using a Protractor* ........................................................ 215

   *Estimating Angles* ......................................................... 219

Types of Angles .................................................................... 219

   *It's an Acute Little Angle* ............................................... 220

   *Don't Be Obtuse* ........................................................... 221

   *The Right Answer* .......................................................... 222

Special Pairs of Angles .......................................................... 222

   *Complementary Angles* ................................................. 223

   *Supplementary Angles* ................................................... 224

Other Types of Angles ........................................................... 225

   *Convex* ......................................................................... 225

   *Concave* ........................................................................ 226

   *Straight* ........................................................................ 227

Parallel and Perpendicular Lines ............................................ 227

**15 Poly(gon) Gets a Cracker** **229**

Classifying Shapes...................................................................229
   *Types of Triangles* .............................................................230
   *Types of Quadrilaterals* .....................................................233
Polygons.................................................................................237
   *Regular Polygons* ..............................................................238
   *Irregular Polygons* ............................................................239
   *Interior Angles* .................................................................239
Perimeter ...............................................................................242
Pythagorean Theorem.............................................................244

**16 Area, Volume, and Surface Area** **247**

Calculating Area.....................................................................248
   *Rectangles*.........................................................................248
   *Parallelograms*...................................................................250
   *Trapezoids* ........................................................................252
   *Triangles* ..........................................................................254
Circles....................................................................................254
   *Easy as Pi*.........................................................................255
   *Getting Around*.................................................................256
   *Calculating Area* ..............................................................257
Shapes in Space .....................................................................257
   *Cubes*................................................................................258
   *Prisms*...............................................................................259
   *Pyramids* ..........................................................................261
   *Cylinders* ..........................................................................262
   *Cones*................................................................................263

**17 Proportions** **265**

When Shapes Are the Same ....................................................266
It's a Shape of a Different Size................................................267
Scale Factors and Resizing......................................................268
   *Using the Scale Factor* .......................................................270
   *Calculating the Scale Factor*...............................................271
Change in Areas.....................................................................276

**Part 6:   Data Analysis**                                                    **281**

  **18   Collecting Data**                                                     **283**

        What Is Data? ...........................................................283
           *Why Bother?* ........................................................*284*
           *Qualitative vs. Quantitative* ....................................*284*
        Recording Data in a Frequency Table...............................285
        Sorting Data ..............................................................287
           *Bins*......................................................................*288*
           *Outliers* ...............................................................*289*
        Displaying Data..........................................................290
           *Pictograph* ...........................................................*290*
           *Histogram* .............................................................*292*
           *Bar Graph*..............................................................*293*
           *Stem-and-Leaf Plot* ................................................*294*
           *Venn Diagrams* ......................................................*295*
           *Line Plot* ..............................................................*297*
        Just an Average Joe......................................................298
           *Mean*......................................................................*298*
           *Median*..................................................................*300*
           *Mode* .....................................................................*301*

  **19   Probably Probability**                                              **303**

        Independent Events......................................................304
           *Coins*.....................................................................*304*
           *Dice*.......................................................................*307*
           *Cards*.....................................................................*310*
           *Chances of Not Happening* .......................................*311*

**Appendixes**

    **A   Glossary**                                                         **313**

    **B   Solutions to Practical Practice Problems**                        **321**

        **Index**                                                           **359**

# Foreword

People often wonder why mathematics is so dry—*just routines and no esthetics, formulae are not poetic.* You might want to take a look and solve the problems in this book. Well done! But, to understand, to explore, to invent—this is mathematics' nature! Problem solving—such adventure in the style of Archimedes. Find your own *Eureka*, read these for Amy's problems, Andy's cautions create just positive emotions …

In ancient Greece, mathematics used to occupy a central place among the arts, and today most people think that it is the art of doing sums. In many pre-algebra textbooks what you learn is often reduced to studying techniques which in the context of fine arts could be compared to clipping stone and mixing paints. This book will prepare you to be real artists in doing mathematics.

Many historical examples (to mention just a few—Gauss, Pascal, Ramanujan) show that mathematicians often mature at a much earlier age than people studying other disciplines, and could make great discoveries even in their teens. Specialists say that to become a real mathematician you *must have hope, faith, and curiosity, and **the prime necessity is curiosity***. This means that you must ask yourself *why*, *how*, and *when*, and these queries are the mainspring that will set you going.

The authors of this book have written it so as to enhance the scientist in you—the problems that would be usually given as developing mainly technical mathematics skills are treated here in a way that would provoke your sense of curiosity and your creativity.

And when you are encouraged to work in the spirit of observing some patterns and making your own discoveries, you could see mathematics in a new way—as an area in which you would not only acquire the technique of doing fast calculations but also habits of creative thinking. This book will prepare you not only for solving problems (for which you have a recipe) but also for problem solving (which is to figure out what to do when you don't know what to do, i.e., to know how to handle situations for which there is no algorithm).

Not for nothing did the great mathematician Godfried Harold Hardy describe mathematicians as people who like painters or poets are makers of patterns, except that theirs are more permanent, being made with *ideas*.

The authors of this book, Amy and Andy, share their love for mathematics with you, the readers, by leading you for a great adventure in the jungle of algebra. You will get their advice about how to act in various challenging situations, when and where to look, but most important, why it is so great to explore and even to invent, and how it is possible to extract order from what, at first glance, seems to be chaos. Enjoy your mathematical adventures!

—Jenny Sendova

# Introduction

Pre-algebra is the culmination of all of the math that average people really use. I want to convince you that this is worth knowing beyond needing to pass a test (and the math in this book should match what's taught in many middle school classrooms in California, Florida, New York, Texas, and other states). This is worth knowing because it's actually helpful for your life. Depending how old you are, you might need these skills now or you might need them in a few years, but they are definitely important skills that you need to know.

Sure, you hear on the news about how everyone needs to know a lot of math to be successful and competitive in our modern world of high-tech careers and all that. And yes, it's true that if you want to be an engineer or a physician or be in any of the other scientific fields, you'll need to know a lot of math. (And in case you're wondering, it's true: most medical schools require a year of calculus and a year of physics as an admissions requirement.) Even studying business requires a fair amount of math.

If you're at a point in your life where you want to keep your options open, then it's best to learn as much math as you can. The higher math required for most of those fields of study is much easier to learn if you have a good background in pre-algebra and algebra. Right now you're facing the first step on what can be a mathematical journey to a lucrative future.

But even if you're not planning on going on to study science or business or advanced psychology (requires statistics) or anything else, the skills that you learn in pre-algebra are going to come in handy.

It is rare that I ever set up equations in my real life. The last time I did it was when I was trying to figure out how many students we had to fit into a large lecture class in order to save money versus running it as regular-size classes. (The solution was that it wouldn't work—in part because we don't have a large enough room. Word problems in math class rarely have restrictions that keep the solutions from working.) I never use one of algebra's most famous formulas, the quadratic formula, except when I'm teaching class. You certainly can find ways to use algebra and higher math in your everyday life if you try hard enough, but unless you're in a career that uses numbers, you can probably get away with avoiding it.

Now, pre-algebra, on the other hand, has the skills and techniques that everyone should know in order to be a successful and productive member of society. And, probably more important, to avoid getting ripped off by other people. If you're trying to do a financial calculation that's more complicated than making change, then you need to know how to do the calculations in pre-algebra. If you ever plan to borrow money or invest money, to negotiate your salary and raises, or to do anything else that's beyond the realm of a simple cash register, then you're going to need to know how to do this. Especially if you're hoping to make your living by reselling things on the Internet, you'll need to know a decent amount of the math in this book.

Beyond the calculations that will come up in all sorts of ways other than finances, pre-algebra also has a bit of geometry and probability and statistics. Again, these are the sorts of things that will come up in your everyday life. If you ever plan a home renovation project or a landscaping project, then knowing about geometry is going to be a great asset. If you're trying to put fencing around a garden, knowing about perimeter is going to help you. When it comes to figuring out how much soil and plants to buy for the garden, you're going to be using area and volume. These fairly basic calculations can help you buy the amount that you need. Sure, you can guess, but you may end up buying a bunch of extra stuff that can't be returned or you might not get enough and need to make another trip. Life-threatening? Certainly not. But irritating and expensive? For sure.

And the data analyses that you'll learn about are the foundations of the probability and statistics that are the language of every medical study and every government report. If you're going to be making decisions based on research and evidence, then you'll need to know what's been done to the data. Even if you don't crunch the numbers yourself, knowing a little bit about data analysis will help you make good decisions and decide which information and advice to trust.

For those of you who are going on to algebra, geometry, trigonometry, precalculus, and beyond, the material in pre-algebra is essential in order to succeed in those subjects. My calculus students end up missing points because they don't remember how to use the distributive law or how to use the rules of exponents. Whether you do those calculations with plain numbers (like in pre-algebra) or in the more complicated realm of algebra, they are the basic skills that you'll need to learn.

While, yes, this material is important, it's also basic enough that anyone can learn it. If you put in the time carefully reading this book (and, ideally, also the text for the course that you're enrolled in—having two points of view really helps cement the ideas) and doing plenty of problems, you will master the material.

There is an old story about how King Ptolemy asked the famous mathematician Euclid for a shortcut to learn math. Euclid replied, "There is no royal road to geometry," meaning that even a king would need to do lots and lots of problems in order to learn the material. With that in mind, this book has a wealth of problems for you to practice with. You should do all of them as well as the problems in the textbook in your course. With a little bit of hard work, anyone who can do basic arithmetic can learn pre-algebra.

## How This Book Is Organized

This book is presented in six parts:

In **Part 1, "The Building Blocks of Math,"** we'll start off our journey together through the world of intermediate mathematics. You'll learn about different types of numbers, how to put them in line, and the laws governing their kingdom. Armed with this knowledge, you'll be ready to take the subject of pre-algebra by storm.

In **Part 2, "Let's All Remain Rational,"** we teach you how to get along with different numbers. Eventually you'll have to stop attacking numbers and start working with them. You might as well start that here, where the numbers are still rational.

**Part 3, "Radical and Exponential Power,"** is a revolution in mathematics. The development of roots and exponents makes it easier for small numbers to grow, dwarfing their once-powerful neighbors. It's often difficult to understand why numbers behave so badly and get out of line, but we're here to help you learn how to simplify your life and theirs. We even teach you how to make money!

In **Part 4, "Self-Expression Through Algebra,"** you'll get to see a glimpse into the beauty and culture of mathematics. We introduce you to variables and their common uses throughout mathematics. You'll be able to fairly distribute quantities across parenthetical expressions, solve equations, and make complicated messes into nice, neat solutions.

In **Part 5, "Ancient Foundations: Geometry,"** you will become as cultured as the ancient Greeks. You will draw dots on a piece of paper in the guise of plotting points on the Cartesian plane. You will doodle and call it graphing equations. You will look at sharp and dull objects and call them angles. You will learn how to tell if two things are the same and how much the same they are. In more magnificent terms, you will learn how to play mathematical games with paper and (colored) pencil.

**Part 6, "Data Analysis,"** is a lesson in the mumbo-jumbo of Wall Street, the news, and gambling. Data analysis is a part of the world as we know it, and you'll learn how to both understand it and make firm statements about the world around us backed up by facts. Whether trying to corner the market on fruit juice or find out what are the chances of going to jail in Monopoly, you'll peek into the secrets of how the world works.

## Extras

To keep you on track, we've included some hints, tips, and tricks that we hope you find helpful. Here are the types of sidebars that you'll encounter as you read this book.

### Kositsky's Cautions

There's a lot to learn in pre-algebra, and there are a lot of errors that almost everyone makes. With these boxes, I'll try to alert you to what these errors are and give you various strategies for avoiding them.

### def·i·ni·tion

Vocabulary is one of the most important parts of mathematics. Clear, precise words help us all understand what we're talking about. Of course, you have to learn these new words first. You can do that here.

### Timely Tips

I've been doing mathematics for dozens of years, and there are a lot of little tricks and helpful suggestions I have up my sleeve. I'll let these dribble out on the paper wherever they are useful.

### Amy's Answers

Sometimes math seems like magic. But just like most illusions it's not so mystical once you get to know the secrets. I'll explain here how to understand the trade of a mathematician.

### Practice Makes Perfect

To truly understand math, you need to get your fingers dirty and work with it yourself. We give you carefully selected problems that will maximize your understanding with a minimum of effort. By doing these, you'll see some of the subtleties and beauty of pre-algebra.

## Acknowledgments

First and foremost, this book wouldn't have been possible if not for the faith Jessica Faust (our agent) placed in us. Both of us wish to express our deep appreciation for piloting us through the treacherous waters of professional publishing.

Adam Schwartz for his laughter and assistance in compiling the glossary. Meghan Gildea and Yangyang Liu for all of their incredible figures and illustrations on top of their moral support. Amy Hancock and Diana Snook for replacing our bloodshot eyes every several days throughout the editorial process.

Allison Gilmore has been a wonderful help as our technical reviewer and cold reader (though she's usually a nice person). Alli graduated from Washington University with a Bachelor's and Master's in mathematics, spent two years as a Rhodes Scholar at

Oxford earning a Master's in sociology, and is now pursuing a Ph.D. in mathematics at Columbia University.

Amy would like to thank Jim Conant for his patience and support and for cooking dinner when I've been too busy writing; my co-workers at UT for patiently picking up the slack when I've been overwhelmed by all my projects; Nedra Stimpfle and the rest of the English Department at Niskayuna High School for believing that I would be a writer some day; and the Center for Excellence in Education for creating the environment that allowed this project to flourish.

Andy would like to thank Greg Hart, Jerry Johnson, Matt Paschke, and Linda Brown Westrick for inspiring my love for mathematics and teaching; Morgan Appleberry, Cory Pender, Michael Spece-Ibanes, Vera te Velde, and Austin Webb for their support, encouragement, and occasional dousing with cold water; and Linda Granger for her motherly advice and editorial support.

## Trademarks

All terms mentioned in this book that are known to be or are suspected of being trademarks or service marks have been appropriately capitalized. Alpha Books and Penguin Group (USA) Inc. cannot attest to the accuracy of this information. Use of a term in this book should not be regarded as affecting the validity of any trademark or service mark.

# Part 1

# The Building Blocks of Math

The section on all the rules and the properties comes at the beginning of all math books. It's here because it's the most important stuff, and you need to know it in order to get any further, so don't skip it thinking that you'll take a shortcut and get ahead. Here you'll be learning about the fussy rules that must be obeyed if you hope to have all your minus signs in the right places and avoid getting points off.

*"So we've got 750 more blocks to go and it takes 13 hours to put one in place…"*

1

# The Whole Story

## In This Chapter

- ◆ Names for several different types of numbers
- ◆ Working with factors and multiples
- ◆ Positive and negative numbers
- ◆ Rules of arithmetic for signed numbers

Numbers are just numbers, right? You've learned about adding, subtracting, multiplying, and dividing them. You've used them for counting, for working with money, and for measuring distance or time. What more would you possibly need to know about them? As you continue your mathematical journey into pre-algebra, your days of dealing with "just numbers" are about to come to an end.

When you study pre-algebra, you start your journey into the mysteries of mathematics. You can think of math as the code of a mystical society, like you might see in *The Da Vinci Code*, where your decryption abilities can rise to calculus or beyond.

Initiation into the world of mathematics requires you, the student, to learn the vocabulary, traditions, incantations, and rules of this ancient art. The rituals of pre-algebra may seem tricky and arcane at first, but they all fit together and give you the skills you need to succeed at algebra and beyond.

Welcome, novice, to your first steps on the path to higher math. By the end of this book, you'll be ready to tackle the challenges of the next level in algebra.

# Natural Numbers

If you've struggled with math in the past, you might think there is nothing *natural* about numbers and the way they work together. Those in the know about math have developed a classification system for numbers based on their properties and how we work with them. Understanding these categories is the first step toward calculating like a pro. The most basic collection of numbers is called the *natural numbers*. The first numbers you learned were probably the natural numbers, those that describe how many objects you can have starting at 1: 1, 2, 3, .... You can have two hands, ten fingers, a dozen cupcakes, one million dollars. All of these quantities are part of the collection of natural numbers. Another important collection of numbers is the *whole numbers*, the natural numbers together with zero. There are no negatives in the collection of whole numbers.

Don't panic! Remember that these numbers are no different from the numbers you've known and loved since you first learned to count. We've just given the collection a fancy new name so we can easily refer to these particular numbers from now on.

All of the rules you learned for adding, subtracting, multiplying, and dividing numbers will stay the same when you are working with the natural numbers. However, you'll have to be careful once we add a few more types of numbers into the mix.

Later in this chapter you'll learn about the integers, a collection of numbers closely related to the natural numbers, and in the next chapter you'll encounter two more important collections of numbers, the rational numbers and the real numbers.

## def•i•ni•tion

The **natural numbers** are all numbers you can reach by adding 1 to itself. For example,

$$2 = 1 + 1, 10 = \underbrace{1+1+1+ \ldots +1}_{10 \text{ times}},$$

and $1,000,000 = \underbrace{1+1+1+ \ldots +1}_{1,000,000 \text{ times}}$,

so 2, 10, and 1,000,000 are all natural numbers. Thus the natural numbers consist of the numbers: 1, 2, 3, ....

## Timely Tips

To mathematicians, "..." has a special meaning always relating to continuation of the same pattern. If "..." is the last item in a list, it means "continuing the same pattern forever." If "..." is between two numbers in a list it means "fill in with the same pattern between the numbers on either side of it." Mathematicians read ... as "dot-dot-dot."

# Factors and Multiples

When you were growing up did you like to take things apart to see how they worked? If so, you might have a bright future in *factoring*. Factoring is the process of breaking a number down into pieces that can be multiplied together to get back the original number.

We'll need to define a few terms (you've probably seen these previously) before we can talk in more detail about factoring.

- *Skip-counting*: Instead of using every number as you count, you skip the same amount of numbers each time. For example, if you skip-counted by six you would get the list: 6, 12, 18, 24, 30, 36, 42, ….

- *Multiple*: One number is a multiple of a second number if the first appears on the list of numbers you get by skip counting by the second number. This means that 24 is a multiple of 6.

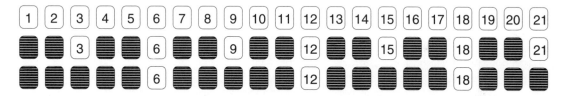

*Skip counting by threes and skip counting by sixes.*

- *Remainder*: If you divide a natural number by another natural number and the second is not a factor of the first, what's left over is called the remainder. For example, if we calculate, 17 ÷ 3 we see that 3 goes into 17 five times and there are 2 left over, so the remainder of 17 ÷ 3 is 2.

- *Factor*: One number is a factor of another if you can divide the second by the first and end up with no remainder.

**Kositsky's Cautions**

Many students confuse factors with multiples. The factor is always the smaller number and the multiple is always the larger number. Since 36 ÷ 12 divides evenly without a remainder, we can say that 12 is a factor of 36 and that 36 is a multiple of 12.

◆ *Divisible*: "Is divisible by" means the same thing as "is a multiple of."

◆ *Prime number*: Any natural number greater than 1 that is divisible only by 1 and itself (and no other natural number) is called a prime number. Numbers like 2, 3, and 5 are prime. So are 61, 727, 7919, and infinitely many other numbers.

### Timely Tips

In Chapter 10, you will discover how to quickly test whether any natural number is prime or not! For now, please memorize the rule that any number between 10 and 100 is prime if it cannot be divided evenly by 2, 3, 5, or 7.

**Example:** Find the factors of 60.

**Solution:** The best way to approach the problem is to start with 1 and keep trying numbers in an orderly way until you've discovered all the factors.

First try 1. That's really easy because every natural number is divisible by 1. So 1 is a factor of 60. It is also easy to see $60 \div 60 = 1$, so 60 is a factor of 60.

Next try 2. You calculate $60 \div 2$, which comes out evenly as 30. This tells you that 2 is a factor of 60.

To save you work later, it also tells you that 30 is a factor of 60 since if $60 \div 2 = 30$, then we also know that $60 \div 30 = 2$.

### Amy's Answers

"It has come out evenly" is a common way of saying that a particular division problem has no remainder.

After trying 2, you can go on to try 3, 4, 5, and 6, discovering in the process that:

$60 \div 3 = 20$

$60 \div 4 = 15$

$60 \div 5 = 12$

$60 \div 6 = 10$

This gives you the following list of factors of 60: 1, 2, 3, 4, 5, 6, 10, 12, 15, 20, 30, and 60. At this point, you might start to suspect that 60 is divisible by everything. Don't be lulled into believing that! When you try $60 \div 7$, you get an answer of 8 with a remainder of 4. Because there is a remainder, 7 is not a factor of 60. As you keep trying numbers, you'll see that neither 8 nor 9 divide evenly into 60, so neither of those are factors, either. At this point, you would want to try the number 10, but because it's already on your list of factors, you already know that 60 is divisible by 10. Once you reach a number that already appears on your list of factors, you're done! Therefore, the complete list of factors of 60 is: 1, 2, 3, 4, 5, 6, 10, 12, 15, 20, 30, 60.

# Prime Factorizations

There are prime cuts of beef, the prime meridian, the prime rate, and, in some countries, a prime minister. All of them are important in some way. You're about to learn about the prime factorization of a natural number.

In the previous section, you found the list of factors of a natural number. Now we're going to find the list of prime numbers that are needed to build a natural number.

**Example:** Find the prime factorization of 60.

**Solution:** The first step is to find any two numbers that multiply to give you 60. Since you're already a champ at working with the factors of 60, this shouldn't be too hard at all. We're going to pick 3 and 20, but you could pick 2 and 30 or 6 and 10 or any other pair of factors that multiply together to give you 60. The only thing that you can't do is pick 1 and 60. This choice gets us nowhere since we have to factor 60 once again!

If any of the factors are primes, then you don't have to break them down any further. Next, continue factoring the composite factors until only primes remain on the bottom-most layer of the tree.

> **Practice Makes Perfect**
>
> Problem 2: Find the factors of 17.
> Problem 3: Find the factors of 48.

**Timely Tips**

Natural numbers greater than 1 that are not prime are called *composite*. The number 1 is special: it is neither prime nor composite.

**Amy's Answers**

This type of diagram is called a factor tree. You can think of the number you're trying to factor as the base of the tree. From the base, branches sprout out (composite numbers) and eventually end with leaves (prime numbers).

On a factor tree, branches of a tree must have other branches or leaves sprouting from them, so you must keep investigating them. When you have no more unexplored branches, you are finished!

Because 3 is a prime, don't do anything to it.

Twenty is composite, and it can be factored into 2 and 10.

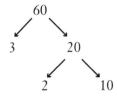

You don't have to do anything else to 2 because it's prime, but you can factor 10 into the product of 2 and 5.

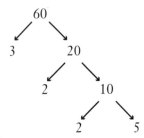

Once you have explored all branches, you're done, and you can write the number as a product of primes (the values in the leaves).

    **Timely Tips** _____

If you want to check your work on a prime factorization problem, all you need to do is make sure that every number on your list is a prime and that when you multiply them all together you get the original number back. If any of these don't work, you need to find your mistake or start the problem over again.

So the prime factorization of 60 is $2 \cdot 2 \cdot 3 \cdot 5$.

Sure enough, you can perform multiplication on the prime factorization and see that $2 \cdot 2 \cdot 3 \cdot 5 = 60$.

## Greatest Common Factor

Have you ever met someone and tried to find out if you knew people in common? Maybe both of you know someone who you went to school with? Perhaps you both had the same tennis coach? When we're talking about numbers, we try to find a factor they both have in common.

**Practice Makes Perfect**

Problem 4: Find the prime factorization of 52.

Problem 5: Find the prime factorization of 38.

Just as people might have a lot of friends in common, numbers also often have many factors in common. In most situations, we're not interested in all the common factors—just the *greatest common factor*. As the name suggests, it's the largest of all the common factors.

## def•i•ni•tion

A **common factor** of two numbers is a third number that evenly divides the other two. For example, 70 and 105 have 7 as a common factor as $70 \div 7 = 10$ and $105 \div 7 = 15$.

The **greatest common factor** of two numbers is a third number that is the largest of the common factors of the other two numbers. For example, 1, 5, 7 and 35 are all common factors of 70 and 105, but only 35 is the greatest common factor.

**Example:** Find the greatest common factor of 48 and 60.

**Solution:** Start by making the list of all the factors of these numbers.

The factors of 48 are 1, 2, 3, 4, 6, 8, 12, 16, 24, and 48.

The factors of 60 are 1, 2, 3, 4, 5, 6, 10, 12, 15, 20, 30, and 60.

By looking at the lists side by side, you can see that the largest number that appears on both lists is 12.

So the greatest common factor of 48 and 60 is 12.

If there are too many factors to keep track of them all in your head, you can write down a list of all the factors and begin crossing out the largest distinct factors. For example, you could start by noticing that neither 60 nor 30 is a factor of 48 and cross them out:

1, 2, 3, 4, 6, 8, 12, 16, 24, and 48

1, 2, 3, 4, 5, 6, 10, 12, 15, 20, ~~30~~, ~~60~~

Continuing, we notice 20 and 15 are not factors of 48, leaving us with:

1, 2, 3, 4, 6, 8, 12, 16, 24, and 48

1, 2, 3, 4, 5, 6, 10, 12, ~~15~~, ~~20~~, ~~30~~, ~~60~~

Since 12 is the largest number on both lists, 12 is the greatest common factor of 60 and 48.

> **Practice Makes Perfect**
>
> Problem 6: Find the greatest common factor of 40 and 60.

> **Practice Makes Perfect**
>
> Problem 7: Use prime factorization to find the greatest common factor of 52 and 60.

Another way to calculate the greatest common factor is by using the prime factorizations of the numbers. Write each number as a product of primes, and find all the primes that they have in common. In order for a prime to be included, it must appear on both lists. If a prime appears more than once in both factorizations, then include it the least number of times that it appears in either of the prime factorizations.

**Example:** Find the greatest common factor of 40 and 48 by using their prime factorizations.

**Solution:** The prime factorizations of 40 and 48 are $40=2 \cdot 2 \cdot 2 \cdot 5$ and $48=2 \cdot 2 \cdot 2 \cdot 2 \cdot 3$.

The only prime factor that is common to both lists is 2.

Two appears three times on the list of factors of 40 and four times in the list of factors of 48. You choose the least number of times it appears, in this case three.

Therefore, the greatest common factor of 40 and 48 is $2 \cdot 2 \cdot 2 = 8$.

# Least Common Multiple

Some ideas come in pairs, like cookies and milk or chocolate and vanilla. Sure, you can have one without the other, but it's the relationship between them that really makes them great. The same is true in math: positive is matched with negative, addition is matched with subtraction, and many other ideas come in pairs. Partnered with the greatest common factor is the least common multiple. One deals with factors, the other with multiples. One looks for the greatest, the other for the least. But both ideas will be used together when you study the arithmetic of fractions.

A *common multiple* of two numbers is a third number that is divisible by the other two. For example, 20 is divisible by both 2 and 5. This means that 20 is a common multiple of 5 and 2.

When you're dealing with numbers that are relatively small (in this case probably less than 50 as opposed to eleventy-squintillion), the easiest way to find the least common multiple is by using skip-counting and making a list.

**Example:** Find the least common multiple of 12 and 18.

**Solution:** We start by making a list of the multiples of 12 and 18 by using skip-counting.

The multiples of 12: 12, 24, 36, 48, 60, 72, …

The multiples of 18: 18, 36, 54, 72, 90, 108, …

Looking at these lists, you should see that there are some multiples in common.

In this case, 36 and 72 appear on both lists.

Because you're looking for the *least* common multiple, you must choose the smallest number that appears on both lists. Therefore, 36 is the least common multiple of 12 and 18.

Just like with the greatest common factor, you can use the prime factorization to find the least common multiple. Both the skip-counting and prime factorization methods should give you the same answer (as long as you do them correctly!)

Once again, you start by finding the prime decomposition of each number. This time you're going to use all the primes that appear on either list. If any prime is repeated in one of the prime factorizations, you must use the prime the same number of times to find the least common multiple. If any prime is repeated on both of the prime factorizations, you must use the prime the maximum number of times it appears on either list.

**Amy's Answers**

There are several ways to calculate greatest common factors and least common multiples. All of them give the same answer if done correctly, so you may pick the method that makes the most sense to you.

**Practice Makes Perfect**

Problem 8: Find the least common multiple of 15 and 20 by skip-counting.

Problem 9: Find the least common multiple of 7 and 12 by skip-counting.

**Kositsky's Cautions**

Prime factorization sometimes masquerades as the prime decomposition. The two are one and the same.

**Example:** Use the prime factorization to find the least common multiple of 20 and 24.

**Solution:** 20=2·2·5 and 24=2·2·2·3. We use every prime that appears in either prime decomposition: 2, 3, and 5.

The primes 3 and 5 each appear only once. Both will appear once in our least common multiple.

The prime 2 appears twice on one list and three times on the other. Since we choose the largest number of times it appears, we must use it three times in our least common multiple.

Therefore, the least common multiple of 20 and 24 is 2·2·2·3·5=120.

---

**Practice Makes Perfect**

Problem 10: Use the prime factorization to find the least common multiple of 10 and 16.

---

# Integers

The next step on the number hierarchy is *integers*. These are numbers that you are at least somewhat familiar with; they just have a new name. Integers are all natural numbers, the negatives of the natural numbers, and zero (no fractions allowed).

**def•i•ni•tion**

Integers are all numbers that can be written without a decimal or fraction.

The integers consist of all numbers that can be reached by adding or subtracting ones from zero. These consist of the natural numbers, the negatives of the natural numbers, and zero, or in numerical form: ..., –3, –2, –1, 0, 1, 2, 3, .... For example,

$$\underbrace{0-1-1-1-...-1}_{10 \text{ times}} = -10,$$ so –10 must be an integer.

Here is another way of looking at the collection of integers:

- There are three types of integers: positive, negative, and zero. There are no others.

- The positive integers such as 1, 2, 3, ... are the same as the natural numbers.

- The negative integers such as –1, –2, –3, ... are the negatives of the natural numbers.

- Zero is also an integer.

# The Number Line and Absolute Value

You can imagine all of the integers living on a single, very long street. Mathematicians call this street the *number line*. You can imagine it stretching infinitely far in both directions. The negative numbers live on the left half of the street, and the positive numbers live on the right half of the street. Zero is the no-man's-land in between the two halves. Right now we're imagining a nice suburban street where the houses have big yards and the numbers have room to stretch out. Later, more numbers will move into our neighborhood, and it will become more crowded than the heart of New York City or downtown Tokyo.

**Kositsky's Cautions**

You might need a new way to visualize numbers when you start working with integers. The natural numbers are fine for counting sheep, but you can't count them as well with integers. You can't have negative sheep!

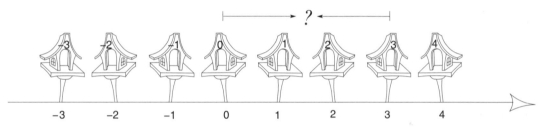

*Our numbers arranged on the number line. How far is it from 0 to 3?*

One reason for imagining the integers as houses on the number line is that it helps you visualize relationships between them. Mathematicians find one particular relationship really valuable: how far away is a number from zero. This relationship is known as a number's *absolute value*.

The notation for absolute value is to surround the number by a set of vertical bars, such as $|17|$ or $|-23|$. The absolute value of a number is one of the easiest things in

**def•i•ni•tion**

A **number line** shows the relationships between all numbers with zero at the center, positive numbers to the right, and negative numbers to the left.

The **absolute value** of a number is its distance from zero on the number line.

math to calculate. Don't confuse yourself by making it harder than it is. It is not difficult once you know how to do it. All you need to do is follow these two simple rules:

If the number is positive or zero, leave it alone. If the number is negative, make it positive.

That's all there is to it. The absolute value leaves positive numbers and zero untouched, and it turns negative numbers into positive numbers. Why does this make sense? Because a number such as –4 is four steps from 0.

**Example:** Calculate $|8|$.

**Solution:** Since 8 is positive, $|8|=8$.

**Example:** Calculate $|-16|$.

**Solution:** Since –16 is negative, $|-16|=16$.

**Example:** Calculate $|0|$.

**Solution:** Since 0 is zero, $|0|=0$.

> **Practice Makes Perfect**
>
> Problem 11: Calculate: $|12|$
>
> Problem 12: Calculate: $|-53|$
>
> Problem 13: Calculate: $|-0|$

# What's Your Sign?

You might say that the integers are the more sophisticated cousins of the natural numbers. When you're calculating in the natural numbers, it doesn't make sense to ask the question, "What is 4 – 7?" However, that's a perfectly acceptable question if you're working with integers.

Instead of imagining numbers as standing for quantities of physical objects such as cats or cars or carts, a good way to think about integers is to imagine that they represent whole amounts of money. The integer 5 would mean that you have or receive $5. On the other hand –5 would mean that you owe someone $5 or that you pay someone $5. Positive integers stand for money that you have or you earn while negative integers stand for money that you owe or you pay. When you add and subtract integers, you can think about it as figuring out how much money you have at the end of a problem (or day).

## Adding and Subtracting

The trickiest part of adding and subtracting integers is figuring out what to do when you subtract a larger number from a smaller number, such as 4 – 7. If you imagine it in terms of money, this would be like a situation where you start out with four dollars and then spend seven. You can't do that without borrowing money, so you'll end up with a negative number representing your debt. If you spend your four dollars, you're left with nothing. As you spend three more dollars (for a total of seven dollars spent), you wind up three bucks in the hole. Keep it up like this, and you'll have to start bringing your lunch from home because you won't have any money left to buy lunch!

What you might also notice is that this is the negative of what you'd get if you switched the order of the numbers you were subtracting: when you calculate 4 – 7, it's the negative of 7 – 4!

In order to visualize what happens when you add or subtract integers, it helps to think about walking down the number line. First you need to find out where to start. This is the value (or address) of the first number. Then find out whether you're walking forward or backward. If you're adding, you're walking forward. If you're subtracting, you walk backward. Next you use the sign of the second number to determine which way you're facing: right if positive, left if negative. Finally, move the proper distance according to the walking rules you found, and you're finished!

**Kositsky's Cautions**

Remember: the order of subtraction matters! When you subtract a bigger number from a smaller number, you'll get a negative number as an answer. The other way around, you'll get a positive answer. Subtracting in the wrong order can turn your million dollar profit into a million dollar debt!

**Timely Tips**

The problem is unchanged if instead of starting with four and taking away seven we imagine that we start with four and add a negative seven. This is because subtracting a positive number is the same as adding a negative number. Whenever you see two signs in a row, if they are the same, you can replace them by a single plus sign. If they are different, you can replace them by a single minus sign.

**Example:** 2 + (–3).

**Solution:** We start our journey at 2.

Since we are adding, we know we must walk forward.

Since –3 is negative, we know we will face left.

We walk: forward, facing left, starting at 2, a distance of three, and we arrive at –1.

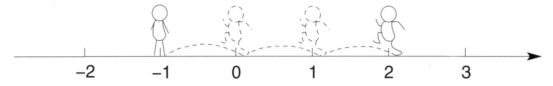

*Computing 2 + (–3) on the number line. Starting at 2, you face left while walking forward three spaces.*

One way to convince yourself adding a negative is the same as subtracting a positive, and subtracting a negative is the same as adding a positive, try imagining each situation on the number line:

♦ You walk forward 2 steps while facing right

♦ You walk backward 2 steps while facing left

♦ You walk forward 2 steps while facing left

♦ You walk backward 2 steps while facing right

In the first two situations you step twice to the right, and in the last two you step twice to the left.

| Practice Makes Perfect |
| --- |
| Problem 14: Calculate: 13 + 5 |
| Problem 15: Calculate: 13 – 5 |
| Problem 16: Calculate: –13 + 5 |
| Problem 17: Calculate: –13 – 5 |
| Problem 18: Calculate: –13 – (–5) |
| Problem 19: Calculate: 13 – (–5) |
| Problem 20: Calculate: 13 + (–5) |
| Problem 21: Calculate: –13 + (–5) |

**Example:** Calculate 8 + (–3).

**Solution:** Since adding a negative number is the same as subtracting a positive, you can rewrite this 8 – 3 = 5.

**Example:** Calculate 7 + (–18).

**Solution:** This is the same as 7 – 18.

When you subtract a larger number from a smaller one, you can compute the answer by changing the order and changing the sign.

Because 18 – 7 = 11, the answer to this question is –11.

# Multiplying and Dividing

Multiplying or dividing signed numbers is much easier than adding or subtracting them. If both numbers are positive, you use the usual rules of multiplication or division and the answer is a positive number. Simple. The next easiest case is when one of the numbers is positive and the other is negative. Do the problem as if neither were signed and make the answer negative. The final case is where both numbers in the problem are negative. Then the answer is a positive number.

**Example:** Calculate 18 ÷ (–3).

**Solution:** We calculate the answer as if there were no signs on the numbers, 18 ÷ 3 = 6.

Since exactly one of the numbers in our division problem is negative, the final answer must be negative.

So 18 ÷ (–3) = –6.

> **Practice Makes Perfect**
>
> Problem 22: Calculate: –13 × (–4)
> Problem 23: Calculate: –9 × 2

## The Least You Need to Know

- ◆ Natural numbers are the counting numbers starting with 1 and continuing forever.

- ◆ The natural numbers, their negatives, and zero combined as a single collection are called the integers.

- ◆ To find the greatest common factor of two numbers, make lists of all of the factors of both numbers and choose the largest one that appears on both lists.

- ◆ To find the least common multiple of two numbers, use skip-counting to make lists of their multiples, and choose the smallest multiple that appears on both lists.

- ◆ Adding a negative is the same as subtracting a positive. Subtracting a negative is the same as adding a positive.

- ◆ When you multiply or divide two integers, if the signs are the same the answer is positive; if the signs are different the answer is negative.

# Filling In the Number Line

## In This Chapter

- ◆ Decimal numbers
- ◆ Fractions
- ◆ Rational numbers
- ◆ Real numbers

The natural numbers are really great. Charming, popular, and very useful for some things. If you're a hunter-gatherer on the savannah and you need to count how much fruit you've collected or how many wildebeests your hunting party has slain, you'd use the natural numbers.

The problem comes when you try to run a modern, market-based economy with them. If you stick with whole numbers, you can't always divide things fairly into parts. If you go out with two other hunters and the three of you only slay one wildebeest, suddenly you have a great interest in the existence of numbers like ⅓. Not to mention that you won't be able to calculate the return on your investment in a company that leads wildebeest hunting expeditions. Unless you always want to risk getting less than your share (or fighting off your friends and family to take the entire wildebeest for yourself), you're going to need to use fractions and decimals.

Many people have a fear of fractions and greatly prefer decimals. They find messy decimals like 0.142857143 ... preferable to work with than elegant fractions like ¹/₇. Both fractions and decimals have a place in our number system. Usually, an answer given by a fraction is exact while one given with a decimal may be an approximation. The one advantage of decimals is that if you're willing to give up some accuracy you can put them in your calculator fairly easily. We introduce you to both fractions and decimals in this chapter; you'll learn all you need to know about fractions in Part 2.

# Getting Down with Decimals

Right now the number line is fairly spacious. The integers have plenty of room to stretch out and relax, but their neighborhood is soon going to fill up with housing developments belonging to decimal numbers.

Decimal numbers are any numbers written with only the digits 0-9 and, at most, one decimal point, no fraction bars, division signs, or other symbols allowed! Some examples of decimal numbers are 4.13, 3.14159, 256, or 173,472,984.7193; decimal numbers such as these are important for filling in the number line.

Even if decimal numbers in general don't make sense to you right now, you're probably familiar with decimal numbers with up to two decimal places in the form of money. The amount of *decimal places* in a number is the quantity of digits to the right of the decimal point.

### Timely Tips

The amount of *decimal places* in a number is the quantity of digits to the right of the decimal point. For example, 2.71828 has five decimal places, 1.6 has one decimal place and 43 has zero decimal places. If you need to count decimal places on a very long number, I recommend you make a small mark beneath every three decimal places, like this: 3.141592653. By using the dots as a counting guide, it is much easier to see this number has 10 decimal places.

## Decimals Make the World Go Round

We all know $4.13 means four dollars and thirteen cents, or an amount of money equal to four one-dollar bills and one dime and three pennies. The first digit to the left of the decimal point is the number of ones (one-dollar bills), the first digit to the *right* of the decimal point is the number of tenths (dimes), and the second digit to the right of the decimal point is the number of hundredths (pennies). The pattern continues with the third digit to the right of the decimal place being thousandths, the fourth digit being ten-thousandths, the fifth hundred-thousandths, and so on.

Let's look at this in a table. Keep in mind that the entries in each row are different ways of saying the same thing.

| Money | Words | Decimal Notation |
|---|---|---|
| 1 \$100 bill = 10 \$10 bills | 1 hundred = 10 tens | $100 = 10 \times 10$ |
| 1 \$10 bill = 10 \$1 bills | 1 ten = 10 ones | $10 = 10 \times 1$ |
| 1 \$1 bill = 10 dimes | 1 one = 10 tenths | $1 = 10 \times 0.1$ |
| 1 dime = 10 pennies | 1 tenth = 10 hundredths | $0.1 = 10 \times 0.01$ |
| -- | 1 hundredth = 10 thousandths | $0.01 = 10 \times 0.001$ |

Different decimal numbers can be closer to each other than different amounts of money can be because there is nothing smaller than a penny in our currency system, but there is a decimal place with a value smaller than a hundredth. However, you can still think about decimal numbers as money as long as you keep in mind that everything to the right of the second decimal place is smaller than one penny.

**Timely Tips**

Reading the parts of numbers to the right of the decimal place is easy once you know how! When you get past the integer part of a decimal number, you say, "point," then read the digits one-by-one, like an American phone number. For example, mathematicians read 316.242 as, "three-hundred-sixteen-point-two-four-two."

## Keeping It Real

The broadest category of numbers you'll see in pre-algebra is the set of *real numbers*. Any number that you could possibly write down using the digits 0-9 and an optional decimal point (if you want to) is a real number. Nice numbers like 23 are real. Somewhat more peculiar numbers like –18, 236, 394 and 268 are real. Special numbers like 0 are real. Fierce decimals like 324.3294843124569803958 are real. So are decimals like 28.44444 …. Specially defined numbers like π are real, too, even though the way to write π as a decimal has an infinite number of digits to the right of the decimal place.

Every number you will encounter in pre-algebra is real. At some point, you may take an algebra course in which you encounter, I kid you not, imaginary numbers. Imaginary numbers are not real.

# Starting Rationally

Nothing sounds like a better idea than *rational numbers*. Sounds like something that should make sense, where there is a logical and coherent basis to working with them. However, there is good news and bad news: Yes, these things are all true.

Rational numbers are fantastic numbers, and the methods for working with them make perfect sense—once you know and understand them. The problem is, you might have gotten off to a bad start with these numbers when you met them and not even realized it.

The rational numbers tend to do most of their work under their stage name. Just like if you heard about "Amanda Lee Rogers," you might not realize that's just another name for the actress Portia de Rossi. Rational numbers are better known as fractions, though technically a fraction is rational only if it can be written with nothing but integers and the fraction bar. We're not sure why the rationals also use an alias. Maybe it's because they've given a bad impression to a lot of people and they want to make a fresh start? Now is a great time to be introduced (or reintroduced) to rational numbers and start a long and fruitful relationship.

Before we dive into fractions, I want to tell you some boring but necessary definitions that will help us understand rational numbers and decimals. A decimal number is called *terminating* if you don't need an ellipsis to write it. In other words, the decimal number stops at some point. A decimal number is called *repeating* if, after some quantity of digits to the right of the decimal place, there is a sequence of digits (not all zero, that's cheating!) that repeats itself.

All terminating or repeating decimals are rational, and in fact, all rational numbers can be written as terminating or repeating decimals; this is one of several ways of defining rational numbers. Another useful way of defining rational numbers is any number that has an integer as a multiple.

## def•i•ni•tion

A **rational number** is any number that has an integer as a multiple. For example, 0.74 is a rational number because $100 \times 0.74 = 74$. Any number that is not rational is *irrational*.

## Introduction to Fractions

Even if you think that you don't know anything at all about fractions, remember that there is one fraction that everyone knows: a half. It's so pervasive that it has an English name in addition to its numerical identity. Maybe it just decided that one-twoth was too confusing and changed its name? There seems to be a lot of that going on with the rational numbers.

Think about what it means when you split something up into halves. You take your basic object, be it a cake or anything else, divide it into two equal pieces, and then take one of them. This sums up the fraction process completely and uses all the important parts of how we write a fraction. One half is written as $\frac{1}{2}$, a symbol consisting of the number 1, a fraction bar, and the number 2. Think back at what we just did to our cake: the fraction bar tells us to divide, so we divided the cake. Into how many pieces? Two. How many halves did we take? One. The fraction has given us complete directions on how to cut the cake into pieces and how much we should take. That's an awful big piece of cake, so maybe we need to work with some other fractions before anyone else realizes that we have a cake and comes around asking for some.

### Kositsky's Cautions

Avoid the pitfall of only imagining fractions in terms of round food. While pizza and pie are both delicious and both can be sliced up to represent fractions, they are most useful when considering fractions smaller than 1. To imagine fractions in general, it helps to imagine something that comes in rectangular strips, like sticks of gum. Rectangular cakes are a good option, too.

# Anatomy of a Fraction

There are a few different ways that we use fractions.

- When we break something into equal parts and then look at only some of those parts.

- When we are given a division problem that has been asked but not yet solved.

- When we set up math problems involving ratios and proportions.

**Timely Tips** _____

Don't worry if you can't remember how to divide a smaller number by a larger number. A calculator can take care of that for you for now.

Right now we're going to focus on the first interpretation: breaking things into equal pieces and looking at some of the pieces. This isn't like when you drop a plate and it shatters into a couple of big pieces and hundreds of tiny shards. When we're working with fractions, we like to imagine the pieces being all the same size with few enough to count.

Fractions are written as two numbers separated by a line, called the fraction bar. One number is written above the bar, and the other number is written below the bar. We could keep calling the top number the top number and the bottom number the bottom number, but they have mathematical names that everyone uses. If you don't use them, everyone in math will think that you're an outsider and that you don't really know what you're talking about. Most of the fraction lingo comes from Latin and these are the terms you need to know.

- **Fraction**. Comes from the Latin word fractus, meaning "to break." When you want to divide or break up an object or a collection of objects, you're probably going to be using a fraction.

- **Denominator**. This is the bottom number in a fraction. It tells you how many pieces you break the whole part into. This word is not related to (world) domination, but rather to denomination, which means "to name."

**Kositsky's Cautions** _____

Division by zero is always forbidden. Because fractions are also division problems, you can never have zero in the denominator of a fraction.

- **Numerator**. This is the top number in a fraction. It's related to a word that means "to count," and it tells you how many pieces you have.

If you had a chocolate bar and broke it into 15 pieces and you kept 7 of them, your share would be represented by the fraction $7/15$. The denominator tells you how many pieces the whole is broken into, and the numerator tells you how many of them you have.

**Timely Tips** _____

Sometimes the fraction bar is written as a horizontal line and sometimes as a diagonal line. Both notations mean the same thing. Which version to use tends to depend on your sense of style and whether you're trying to type or write the fraction easily. The slanted bar / is also used for division; because fractions are also division problems, this is okay.

# Lining Up on the Number Line

When we were introduced to the integers and the number line back in Chapter 1, there was plenty of room for all the numbers to fit on the line and to have some space to spread out. Now that we have decimals and fractions, the number line is going to be more crowded than Disney World during a school vacation. Every real number has an assigned space on the number line corresponding to its value. Even if we write a number in a different format, its place on the number line remains unchanged.

## Numerical Comparison

There are six primary ways to describe how numbers are situated by each other. You've probably seen these before, so we'll only introduce them briefly and give you a few practice problems.

- **Equal, =.** This relationship means that two numbers share the same spot on the number line. For example, $\frac{1}{2} = 0.5$.

- **Not equal, ≠.** Two numbers are not equal to each other when they inhabit different spots on the number line. For example, $7 \neq 3$.

- **Less than, <.** The first number is less than the second number if it is to the left of the second on the number line. For example, $-2 < 1$.

- **Less than or equal to, ≤.** The first number is less than or equal to the second number if it is to the left of the second on the number line or if the two numbers are equal. For example, $-1 \leq -1$ and $-2 \leq 4$.

- **Greater than, >.** The first number is greater than the second number if it is to the right of the second on the number line. For example, $-7 > -23$.

- **Greater than or equal to, ≥.** The first number is greater than or equal to the second number if it is to the right of the second on the number line or if the two numbers are equal. For example, $3 \geq 0$ and $0 \geq 0$.

# Decimals on the Number Line

Placing a decimal number on the number line sometimes takes a few steps. The easiest case is when your decimal number is an integer, such as –4 or 0 or 2. (To remember that they're also decimals, you might want to write them as –4.0, 0.0, and 2.0. They're still the same numbers.) If that's the case, you merely plot it on the number line that you're used to.

But what if you're trying to put a number like 2.3 or –1.2 on the number line? The first thing you do is to subdivide the spaces on the number line into 10 smaller spaces. Usually this is easier if you start by drawing your number line fairly large. To put 2.3 on the number line, you mark it at the third tick mark beyond 2. Putting –1.2 is a little bit trickier because it's a negative number. Since –1.2 is more negative than –1, it must be placed to the left of –1 on the number line. Put –1.2 on the number line, two tick marks to the left of –1.

## Kositsky's Cautions

Always remember to go to the left as the absolute value of the negative number increases. For example, |–1.2| > |–1.0| (the absolute value of negative 1.2 is greater than the absolute value of negative one), so –1.2 is to the left of –1.0 on the number line.

## Timely Tips

The reason that we keep subdividing the intervals into 10 pieces is because we work in a base-10 number system. The digits to the left of the decimal point tell us how many groups of 10 we have. Place values to the right of the decimal point tell us how many pieces we have after subdividing by 10.

What if you have more decimal places? The practical answer is that at this point you begin to fudge it. But the official answer is a little bit more complicated.

Fudgery, aside from being quite tasty, has many applications and some rather precise rules. When you're asked to round a decimal number to the nearest tenth, it's customary to fudge your number up if your number has a hundredths value of 5 or greater, and to round down if it's hundredths value is 4 or less. This same pattern continues for rounding to the nearest integer, the nearest hundredth, the nearest thousandth, and so on. For example, 9.3459 rounded to the nearest tenth is 9.3, to the nearest hundredth is 9.35, to the nearest thousandth is 9.346.

**Example:** Place 2.47 on the number line.

**Solution:** In order to do a problem like this, you'll need to be very careful how you draw your number line.

You'll take the region between 2 and 3 and break it into 10 pieces.

Then you'll take the region between the fourth and fifth tick marks and break that into 10 pieces. Then you can put 2.47 on the number line.

If you were given a number with more decimal places, you would just keep continuing this process. In reality, it's unlikely that anyone would ask you to subdivide the number line into a thousand parts just to mark the location of a number with three digits after the decimal point.

> **Practice Makes Perfect**
>
> Problem 2: Place the following numbers on the number line: −2.6, −1.2, 0.7, 1.25, 3.2.

## Fractions on the Number Line

Placing a fraction on the number line is usually pretty easy. The basic case is when you're working with a positive fraction where the numerator is smaller number than the denominator. Then the denominator tells you how many spaces to divide the interval between 0 and 1 into. The numerator tells you how many to count over.

**Example:** Put the fraction $^3/_7$ on the number line.

**Solution:** Divide the interval between 0 and 1 into seven equal pieces. Start at 0 and count over 3 to the right of them. That's where $^3/_7$ lives.

**Example:** Place the fraction $-^2/_5$ on the number line.

**Solution:** Since this is a negative fraction, it will go to the left of zero. Divide the part of the number line between 0 and 1 into five pieces (because the denominator of the fraction is 5). Then count over two of them to the left of zero. That's where $-^2/_5$ lives on the number line.

It gets a little bit more complicated if the numerator is a larger number than the denominator, but it's nothing that you can't handle. You might not be used to working with this type of fraction, but they are perfectly fine numbers. Just follow these steps, and you'll be able to locate your fraction on the number line. We'll assume that your fraction is positive.

 **Timely Tips**

American rulers are sections of the number line divided up into $^1/_{16}$ or $^1/_8$ inch sections. Metric rulers are sections of the number line divided up into $^1/_{1,000}$ meter sections.

1. Divide the numerator by the denominator. If you get a whole number, you're done and can plot it on the number line.

2. If there was a remainder and your fraction is positive, go up to the next whole number, and call this the "stopping number."

3. Using the denominator of your fraction, divide the space between 0 and 1 into that many pieces.

4. Do the same thing with the space between 1 and 2.

5. Continue until you reach the stopping number.

6. The numerator of your fraction tells you how many spaces to count up. Start at zero and count up that many.

**Example:** Place the fraction ¹⁷⁄₃ on the number line.

**Solution:** Following the steps.

1. If you calculate $17 \div 3$, you'll get 5 with a remainder.

2. Since there is a remainder, the stopping number is 6.

3. Break the space between 0 and 1 into three pieces, because 3 is the denominator of the fraction.

4. Break the space between 1 and 2 into three pieces.

5. Continue with the space between 2 and 3, between 3 and 4, between 4 and 5, and between 5 and 6. Since 6 is the stopping number, you can stop.

6. The numerator of the fraction is 17, so, starting at 0, count up 17 spaces, showing where ¹⁷⁄₃ lives on the number line.

**Timely Tips** _____

Remember that the answer to a division problem can be written either as a quotient and a remainder or as a fraction.

What if your fraction is negative? In that case, the stopping number is the next negative number left from the answer of your division. You'll subdivide the regions to the left of zero until you reach the stopping number, and you'll use the numerator to count spaces to the left.

# Rational vs. Irrational Numbers

We've been using fraction and rational numbers somewhat interchangeably. Technically, we shouldn't be doing that. It's not entirely correct. It's not that we've been lying to you, it's just that we neglected to mention picky details with fractions.

There's a trade-off in mathematics: stating things simply means we have to leave out some of the precision, while stating things precisely means that they are no longer simple.

To explain the official definition of rational numbers, we use the division interpretation of fractions, that a fraction is a division problem where you divide the numerator by the denominator. In this setting the fraction $\frac{1}{3}$ is $1 \div 3$. The fraction $\frac{8}{2}$ is $8 \div 2$, and so on. Using this interpretation, we can restate the definition of rational number in a more clear form.

> **Practice Makes Perfect**
>
> Problem 3: Place the following fractions on the number line: $-\frac{7}{2}$, $-\frac{1}{3}$, $\frac{3}{4}$, $\frac{7}{5}$, $\frac{18}{7}$.

A rational number is a number that meets any of these criteria:

♦ All integers (positive, negative, and zero) are rational numbers.

♦ Any number that can be written as an integer divided by another integer is a rational number. The sign doesn't matter—just don't divide by zero.

♦ Any number that can be written as a terminating or repeating decimal is a rational number.

If a number can be written in any of those ways, then it is a rational number. In Part 2, you'll learn a lot more about working with both rational numbers and fractions.

So what is the big deal? Why make a fuss about which types of fractions get the name "rational" and carefully excluding others? This is part of mathematicians' system for classifying numbers. Maybe you've had a collection of something. Rocks? Tropical fish? Baseball cards? Did you ever sort your collection? Mathematicians sort and classify collections of numbers. You'd think that we'd have better things to do with our time, but we don't. And we expect our students to be able to sort numbers with the best of them!

But wait for it. It gets weirder. Because a fraction is just a division problem, you can turn a fraction into a decimal by doing the division (a calculator is often an excellent tool to accomplish this if you're feeling lazy). Every rational number is also a decimal. "So what?" you might be thinking. It turns out that it doesn't work in the other direction. There are decimals that are not rational numbers! There are decimals that cannot be written as an integer divided by another integer. It is impossible to write them in that form, no matter how hard you try and how sneaky you are. It just can't be done. Numbers that aren't rational are called *irrational*. Every real number is either rational or irrational.

> **Kositsky's Cautions** _____
>
> Don't forget decimals and fractions are different ways of writing numbers whereas a number is rational or irrational regardless of how it is written. Let's look at one-half. 0.5 is a rational number written in decimal notation (it is a terminating decimal) and ½ is a rational number written in fraction notation (½ is written using only integers and a fraction bar). No matter how we write it, one-half will always be rational.

You might be wondering if anyone really cares about this. When the ancient Greeks were figuring out the basics of mathematics thousands of years ago, they were really shocked to learn about irrational numbers. There was a secret society of mathematicians called The Pythagoreans, after their leader, Pythagoras. You learn about their famous theorem regarding triangles in Part 5. Their entire worldview was based on the theory that every number was rational, and we're told that they were shocked when they discovered that this wasn't the case. Their opinions about these numbers persist in the language that we use to describe them: who'd want to be an irrational number? One of the guys in their society told people about this disturbing discovery, so they took him out on a boat, threw him overboard, and he drowned!

Talking about irrational numbers is no longer hazardous to your health, and you'll get to learn a lot more about them later in this book.

## The Least You Need to Know

- Any number you can write using our decimal system is a real number.
- The top number in a fraction is the numerator, and the bottom number in a fraction is the denominator. They are separated by a fraction bar.
- Any two numbers that have the same location on the number line are equal.
- Real numbers come in two varieties, rational and irrational.
- Rational numbers can be written as integer fractions.
- It is impossible to write an irrational number as an integer fraction.

# Properties of Number Systems

## In This Chapter

- ◆ Finding the correct order
- ◆ Working with the laws of the basic operations
- ◆ Using inverses to undo operations

Years ago my little brother interviewed for a job with a fireworks contractor. It was the type of company that cities hire to plan and run the Fourth of July fireworks. They put on fairly elaborate pyrotechnic displays. In some sense this seemed like the perfect job for my brother. He had a fascination with fire throughout his childhood.

My mother was worried, though, because although he had a genuine interest in pyrotechnics, this line of work is fairly dangerous. My brother explained that everyone at the fireworks company was very familiar with the rules of safely handling pyrotechnics and followed the safety rules very carefully. In the end, he didn't get the job. Instead he became a pilot for a skydiving company—another job where it's important to stick to the rules.

Math has rules, too, and it is essential to follow them when working on problems. You might not be risking life and limb, but if you stray from the rules—even a little—you often get the entire problem wrong.

# Order of Operations

Most of the math problems you encounter at this level have one right answer, and your teacher will expect you to get the same answer. When faced with a problem such as 4 + 7 × 3, we need to agree on how to do it: do we add 4 + 7, then multiply the answer by 3, or do we multiply 7 × 3, then add this to 4? It turns out that we must multiply first, then add. If you got 25 as the answer, then you already know something about the *order of operations*, or the set of rules mathematicians use to describe how to properly answer problems such as 4 + 7 × 3. If you did not get 25 as the answer to 4 + 7 × 3, read this section very closely.

## Grouping Symbols

The first rule in simplifying an expression is to always deal with the parts that are inside grouping symbols such as (parentheses), {brackets}, or [braces] before doing anything else. If these are nested, you need to start with the innermost set first.

**Kositsky's Cautions** _____

When you learn more about fractions in Part 2, you will find out that you should imagine a set of invisible parentheses around the top of a fraction and another set around the bottom of a fraction. Invisible parentheses? Is this math or wizardry? You should feel free to grab a pencil and write them in so that you can see them.

## def•i•ni•tion _____

The **order of operations** is a set of rules describing how you should solve math problems with more than one operation. That is to say, the order of operations tells you how to correctly multiply, divide, add, and subtract in problems with more than one times symbol, division symbol, plus sign, or minus sign.

**Example:** Calculate (3 − 8) × [7 × (−2 + 8)].

**Solution:** Because the part (−2 + 8) is grouped together with parentheses and is inside another set of grouping symbols, it is the innermost group, so you should start with it. Because −2 + 8 = 6, the problem becomes:

(3 − 8) × [7 × 6].

Now there are two sets of grouping symbols, the parentheses and the brackets. Both of them are at the same level; they are not nested one inside the other. This means that you can choose which one to deal with first.

Calculating (3 − 8) = −5, and [7 × 6] = 42, so, substituting back in, the problem is now −5 × 42, which equals −210.

# The Next Order of Business

Not all problems have parentheses in them to guide you step-by-step down the path to the right answer. Sometimes you're faced with expressions of the form $5 - 3 \times 7$, and you might not be sure where to start. Fortunately, there are hard-and-fast rules that you follow every time. There are parts of math where you can show your creative and independent spirit and do things your own way. This is not one of them.

> ### Practice Makes Perfect
>
> Problem 1: Evaluate the following expressions.
> (a) $4 + \{8 \times [7 + (2-4)]\}$
> (b) $4 + \{8 \times [(7 + 2)-4]\}$
> (c) $4 + \{[8 \times (7 + 2)]-4\}$

Once you've taken care of all the operations that are written inside of parentheses and other grouping symbols, you next want to do all the multiplications and divisions in the problem, working from left to right.

After you've dealt with grouping symbols, multiplications, and divisions, what's left? Addition and subtraction. In a complicated expression with several operations, there are only two situations where you are allowed to add or subtract. One, if addition and subtraction are the only operation inside a set of parentheses. Or two, if there are no other operations left in the problem.

**Example:** Calculate $5 - 3 \times 7$.

**Solution:** You must multiply before you're allowed to add, so start by calculating $3 \times 7 = 21$. The problem is now $5 - 21$, so the answer is $-16$.

**Example:** Calculate $[16 - 2 \times (14 - 3)] + (-13 - 8) \times \{[(8 \times 3 \div 2) - 2] + [-3 \times 6]\}$.

**Solution:** Wow, that's a doozy. Don't be overwhelmed. Just remember the rules. Start with the innermost parentheses and work your way out. $(14 - 3) = 11$ and $(-13 - 8) = -21$, so those are simple. So is $(-3 \times 6) = -18$. You might need to think about $(8 \times 3 \div 2)$. Remember, if you have just multiplications and divisions, you must work from left to right: $8 \times 3 = 24$ and $24 \div 2 = 12$.

Substituting in the progress you've made, the problem now looks like: $(16 - 2 \times 11) + -21 \times [(12 - 2) + -18]$.

It's like peeling an onion: after you finish one layer, you start on the next. Evaluating $(12 - 2)$, you get 10 and substitute that in: $(16 - 2 \times 11) + -21 \times (10 + -18)$.

You're making progress. The next step is to deal with $(10 + -18) = -8$ and substitute again: $(16 - 2 \times 11) + -21 \times -8$.

One more set of parentheses to deal with, but this time containing two operations: a subtraction and a multiplication. Which do you do first? Do you start by multiplying? Do you begin by subtracting? Do you try to avoid making a decision by scribbling something half-legible in hopes of tricking the teacher into assuming that you've made the right choice?

The order of operations gives us a clear directive: start with the multiplication. A new question rears its head. Do you multiply by 2 or by –2?

It depends on how you're thinking of the problem. Do you see that minus sign as subtracting? Or are you adding a negative? (Mathematically, they're the exact same thing.) If you read it as "sixteen [pause] minus [pause] two-times-eleven," you'd solve it like this: $(16 - (2 \times 11)) = (16 + (-22)) = 16 - 22 = -6$.

But if you read it as "sixteen [pause] plus a [pause] negative-two-times-eleven," you'd do it this way: $(16 + (-2 \times 11)) = (16 + -22) = -6$. In either case, you get –6.

After we calculate $-21 \times -8 = 168$, the expression becomes $-6 + 168$, which is 162.

---

### Practice Makes Perfect

Problem 2: Evaluate the following expressions.

(a) $(16 - 2 \times (14 - 3)) + (-13 - 8) \times (((8 \times 3 \div 2) - 2) + (-3 \times 6))$

(b) $(16 - (2 \times 14 - 3)) + (-13 - (8 \times ((8 \times 3 \div 2) - 2) + (-3 \times 6)))$

(c) $(16 - 2 \times (14 - 3)) + (-13 - 8) \times ((8 \times 3 \div 2) - (2 + (-3 \times 6)))$

(d) $16 - (2 \times (14 - (3 + (-13 - (8 \times (18 \times 3) \div (2 - (2 + (-3 \times 6))))))))$

---

# Commutative Law

The lives of numbers are different from our own. When we talk about an "operation" on a number, it's not having surgery; we're only talking about +, –, ×, ÷, and a few other symbols you'll be introduced to later. When we break the laws of math, no one goes to jail, but the answer may be wrong! And when numbers commute, they aren't headed to work.

The *Commutative Law* is a rule of arithmetic and algebra that says order doesn't matter when adding or multiplying in our number system; two numbers added or multiplied together can switch places without changing the answer. The Commutative Law of Addition states that the order in which you add two quantities does not affect the sum. The Commutative Law of Multiplication states that the order in which you multiply two quantities does not affect the sum. These laws apply to all of the types of numbers that we've seen so far.

What does the Commutative Law of Addition mean in a practical sense? It means that the answer to 2 + 4 will always be same as the answer to 4 + 2. This sort of example may seem very simple, and you might be asking "So what?" but the Commutative Law also can be used to make it easier to wrap your mind around addition of integers.

**Example:** Calculate (–15) + 63.

**Solution:** It's easier to think about this problem after re-writing it using the Commutative Law. Changing the order of the terms, we get 63 + (–15). When we're adding integers, adding a negative number is the same as subtracting a positive number, so this further simplifies to 63 – 15, which equals 48.

**Example:** Calculate 8742.3 + 3499.7 and 3499.7 + 8742.3.

**Solution:**

$$
\begin{array}{r} {}^{3}\\ 8742.3 \\ +3499.7 \\ \hline .0 \end{array}
\quad
\begin{array}{r} {}^{5\ 3}\\ 8742.3 \\ +3499.7 \\ \hline 2.0 \end{array}
\quad
\begin{array}{r} {}^{8\ 5\ 3}\\ 8742.3 \\ +3499.7 \\ \hline 42.0 \end{array}
\quad
\begin{array}{r} {}^{9\ 8\ 5\ 3}\\ 8742.3 \\ +3499.7 \\ \hline 242.0 \end{array}
\quad
\begin{array}{r} {}^{9\ 8\ 5\ 3}\\ 8742.3 \\ +3499.7 \\ \hline 12242.0 \end{array}
$$

and

$$
\begin{array}{r} {}^{10}\\ 3499.7 \\ +8742.3 \\ \hline .0 \end{array}
\quad
\begin{array}{r} {}^{10\ 10}\\ 3499.7 \\ +8742.3 \\ \hline 2.0 \end{array}
\quad
\begin{array}{r} {}^{5\ 10\ 10}\\ 3499.7 \\ +8742.3 \\ \hline 42.0 \end{array}
\quad
\begin{array}{r} {}^{4\ 5\ 10\ 10}\\ 3499.7 \\ +8742.3 \\ \hline 242.0 \end{array}
\quad
\begin{array}{r} {}^{4\ 5\ 10\ 10}\\ 3499.7 \\ +8742.3 \\ \hline 12242.0 \end{array}
$$

The Commutative Law also applies to multiplication. If you have three rows with five apples each, you'll have the same number of apples if you had arranged them into five rows with three apples each. (If you're too lazy to rearrange your apples, you can just hold your head sideways.) In symbols, this idea would be written as $3 \times 5 = 5 \times 3$. This is why there is so much repetition in the multiplication facts, making it less of a burden to learn your times tables.

**Kositsky's Cautions**

When you commute (change the order of) two numbers, you must keep the negative sign with its number! For example, $-5 + 3 = 3 + (-5) = 3 - 5 = -2$, but $-5 + 3 \neq -3 + 5 = 2$.

Don't get complacent and start using the Commutative Law all willy-nilly. It doesn't work for subtraction or division. The problem 4 – 7 is different from 7 – 4. Changing the order of subtraction will change your answer from negative to positive (or vice versa). Subtraction problems have one right answer, and changing the order will almost always give you the wrong answer. (For extra credit, when will changing the order of subtraction *not* give you a different answer?)

If you want to use the Commutative Law to make it easier to work on a subtraction problem, you must first convert it into an addition problem. You might wonder how this is possible; it can be done because subtracting a number is the same as adding its opposite.

**Example:** Convert 4 – 7 to an addition problem and then apply the Commutative Law.

**Solution:** Using the rules of arithmetic for integers, subtracting 7 is the same as adding (–7), so you can rewrite the problem as 4 + (–7).

Now that it's an addition problem, you can use the Commutative Law and change it to (–7) + 4.

You have to be careful with division problems, too. If you are asked to calculate 24 ÷ 6, you'd get an answer of 4. However, if you tried to switch the order and calculate 6 ÷ 24, you'd end up with 0.25 or ¼.

The Commutative Law can be written in mathematical notation. The law for addition might look like this: a + b = b + a, and the law for multiplication might look like this: a × b = b × a. Don't pay too much attention to the letters; they're just placeholders for any number. You'll learn more about these in Chapter 10. Both of these mathematical statements are saying the same thing: when you add or multiply, it doesn't matter which number comes first and which one comes second.

> **Practice Makes Perfect**
>
> Problem 3: For each pair of expressions, determine whether they are the same or different.
>
>   (a) 8 + 4 and 4 + 8
>   (b) 10 – 7 and 7 – 10
>   (c) 4 × 6 and 6 × 4

# Associative Law

Do you have neighbors or a group who you associate with? Numbers do when they're being added or multiplied. The *Associative Law* is a mathematical way of describing how numbers in a math problem can be grouped together in different ways without changing the answer to the problem. The Associative Law of Addition says regardless of the grouping in which you add the numbers in an equation, you will always come out with the same answer. The Associative Law of Multiplication says that regardless of the grouping in which you multiply the numbers in an equation, you will always get the same answer.

Specifically, this means that if you're given a problem like 17 + 24 + 26, you can choose whether to start with the first two numbers or with the second two numbers. You could begin by adding together 17 + 24 to get 41 and then adding 41 + 26 to get 67. Or you could add together 24 + 26 to get 50 and then add 17 + 50 to get 67. When you add together three numbers, you can choose whether to add together the first two and then add in the third or to add the second two and add the first to that result. Either way you group the numbers together, you'll get the same answer.

**def•i•ni•tion** ____

The **Associative Law of Addition** says that the grouping of additions does not matter. For example, (3 + 4) + 7 = 3 + (4 + 7).

The **Associative Law of Multiplication** says that the grouping of multiplications does not matter. For example, (3 × 4) × 7 = 3 × (4 × 7).

Using parentheses as grouping symbols, we can express these two ways of doing the problem as (17 + 24) + 26 and 17 + (24 + 26). When we write this in mathematical notation, we say (a + b) + c = a + (b + c). Again, the letters are just placeholders for any number.

Just like the Commutative Law, the Associative Law will apply when you're multiplying. Using mathematical notation, we would express this law as a × (b × c) = (a × b) × c.

**Example:** Calculate 17 × 5 × 2.

**Solution:** If you have a calculator handy, you probably wouldn't give this problem too much thought: you'd just type in 17 × 5 × 2 = and out would pop the right answer. But what if you didn't have a calculator with you?

It would be easiest to do this problem as 17(5 × 2) = 17 × 10 = 170 rather than trying to calculate (17 × 5) first!

**Timely Tips** ____

Once you are skilled at using the order of operations, you sometimes can save yourself work by making careful use of the Associative Law. You can only use the Associative Law if all the operations in a step are written as addition or if all the operations in a step are written as multiplication.

You still need to be careful if you're working on a subtraction problem or a division problem. In a problem like 8 − 13 − 7, you need to perform the subtraction from left to right: start by subtracting 8 − 13, getting −5, and then subtract −5 − 7, for the final answer of −12. Any method that gives you a different answer is wrong. Unlike with the Commutative Law, there isn't an easy way to use other properties to get around this difficulty. It can be done, but it requires one of the most difficult laws of all: the Distributive Law.

---

| **Practice Makes Perfect** |
| --- |

Problem 4: For each pair of expressions, determine whether or not they are equal.
   (a) (7 + 8) + 5 and 7 + (8 + 5)
   (b) (24 − 5) − 12 and 24 − (5 − 12)
   (c) (2 × 4) × 5 and 2 × (4 × 5)
   (d) (14 + 6) × 7 and 14 + (6 × 7)

# Distributive Law

You probably aren't afraid of the Distributive Law, which is a good thing because there is no reason to be scared of it. However, it might be a good idea to be at least a little bit wary of the Distributive Law, to always keep an eye on it, and to show it some respect. You might want to imagine a movie about organized crime and a powerful character called "The Distributor." Just as long as you deal honestly and properly with The Distributor, you'll be fine.

**Kositsky's Cautions**

One of the major sources of wrong answers in algebra class is a lack of familiarity with the Distributive Law. Some of the problems in this section may seem contrived, but they're designed to help you master this essential skill.

## Standard Problems

Up to this point, the laws have been able to tackle only one operation at a time. All of this changes with the Distributive Law. It is the rule that governs how to solve many problems that involve both addition and multiplication. In expressions where you have numbers being added within parentheses and then multiplied by another term, you can multiply each of the inner terms by the outside number before adding them. For example, $4 \times (5 + 9) = 4 \times 5 + 4 \times 9$. Using mathematical notation (having letters take the place of specific numbers), the Distributive Law is written as $a \times (b + c) = a \times b + a \times c$. When we apply this law to both multiplication and subtraction, it's written as $a \times (b - c) = a \times b - a \times c$. We'll introduce you to these letters properly in Part 3.

When you're working with natural numbers, the Distributive Law isn't particularly difficult.

### Timely Tips

There are three main ways to write multiplication in math, and they all mean the same thing! You can use the familiar ×, you can use a dot written like this ·, or you can sometimes even leave the symbol out altogether! For example, $3 \times (4 \times 5)$, $3 \cdot (4 \cdot 5)$, and $3(4 \times 5)$ all mean exactly the same thing. A good rule-of-thumb is if there is a letter or a parenthesis, you can leave the multiplication symbol out.

**Example:** Use the Distributive Law to rewrite $4 \times (3 + 8)$.

**Solution:** Because you're not asked for the numerical answer, no matter how tempting it is, you cannot give the number 44 as the solution to this problem. Instead, you write $4 \times 3 + 4 \times 8$.

**Example:** Use the Distributive Law to rewrite $5 \times (6 - 8)$.

**Solution:** $5 \times 6 - 5 \times 8$.

> ### Practice Makes Perfect
>
> Problem 5: Calculate $6 \times (8 + 9)$.
>
> Problem 6: Calculate $11 \times (8 - 4)$.
>
> Problem 7: Calculate $(12 - 7) \times 5$.

## Distributing a Minus Sign

You need to be careful with the Distributive Law when multiplying by a negative integer. This is the most insidious appearance of the Distributive Law. This is the source of sign errors and other tales of woe told by math students worldwide. If you can master this skill, you'll breeze through problems that stop other students in their tracks.

**Example:** Use the Distributive Law to rewrite $-7 \times (6 + 8)$.

**Solution:** If you don't use any parentheses as you apply the rule, you'll get $-7 \times 6 + -7 \times 8$, which mostly looks like a mess. While technically correct, it's ugly enough that you might lose some style points.

For a problem like this one, you're probably going to be better off using the parentheses notation for multiplication and starting out by writing $(-7)(6) + (-7)(8)$.

### Kositsky's Cautions

Remember that $-7 \times (6 + 8)$ is mathematically the same statement as $-7 \cdot (6 + 8)$ and $-7(6 + 8)$!

This answer is correct and is good enough for most purposes. At this point, you could consider yourself finished with the problem and move on to the next one.

If you want to wrangle with signs and write the answer in an even more aesthetically pleasing way, notice that the term that you're adding is a negative number, –7, multiplied by a positive number, 8. What you're really doing is adding a negative—which is the same as subtracting a positive.

Therefore, your answer can be somewhat simplified to $(-7)(6) - (7)(8)$. This can be further simplified to $-(7)(6) - (7)(8)$. See what we mean about signs being tricky with the Distributive Law? Since we've already done more than the problem asked for, we're going to leave well enough alone and not do any more to these numbers.

**Timely Tips** _____

Multiplying a number by –1 will change that number's sign. If you're faced with a negative number, you can rewrite it as –1 times its opposite. This means that $-5 = (-1)(5)$. Sometimes this is written as $-5 = -(5)$. This also can be called factoring out a sign.

**Example:** Use the Distributive Law to rewrite $-5 \times (4 - 12)$, then use rules of integers to simplify the signs as much as possible.

**Solution:** There are two ways of thinking about this problem, depending on whether you think about the part in parentheses as $(4 - 12)$ or as $(4 + (-12))$, using the fact that subtracting a positive number is the same as adding a negative number. You might wonder what's up: this looks like complicating the problem instead of simplifying it!

The truth is that while there may be one right answer to many of the problems that you will see in pre-algebra, there are often many different ways to reach that answer.

If you've decided that the part in parentheses should be thought of as $(4 - 12)$, when you apply the Distributive Law, you will get $(-5)(4) - (-5)(12)$. If you don't understand why the sign in the middle is a minus sign, then you might want to consider the other method.

Looking at the signs, $(-5)(4)$, is a negative number and so is $(-5)(12)$. So overall you are taking a negative number and subtracting another negative number from it. Remembering that subtracting a negative is the same as adding a positive, we can finish the problem as $-(5)(4) + (5)(12)$.

What if you had chosen the other path, and you decided to try the problem as $-5 \times (4 + (-12))$? Again, apply the Distributive Law: you'll get $(-5)(4) + (-5)(-12)$. Simplifying the signs results in the same answer as before: $-(5)(4) + (5)(12)$.

This last example is not as simple as it looks. This is the case that confuses the most students, so we've saved it for last.

**Example:** Use the Distributive Law to rewrite –(7 + 10).

**Solution:** I know that you know that the expression in the example happens to equal –17. Doesn't matter.

The answer to this problem is not –17 because you're not being asked what the value is; you're being asked to apply the Distributive Law!

You might not recognize this as being the same type of problem that we've been doing because the notation doesn't make it clear that there is multiplication going on. What's being multiplied is (–1) × (7 + 10). You will almost never see it written this way; in math class this is nearly always written as –(7 + 10).

Now that you see that this is another example of the Distributive Law, you can just apply the rule: (–1)(7) + (–1)(10). Because multiplying a number by –1 changes its sign, we could also write this as –7 + –10.

**Example:** Use the Distributive Law to rewrite –(–3 – 5).

**Solution:** You could think of this as (–1) × (–3 + (–5)). Applying the Distributive Law, you get (–1)(–3) + (–1)(–5) = 3 + 5.

> **Practice Makes Perfect**
>
> Problem 8: Use the Distributive Law to rewrite each expression. Then use the rules of integers to simplify signs, if appropriate.
>
> (a) –6 × (4 + 5)
> (b) –(–11 – 7)
> (c) (6 – 5) × (–9)
> (d) –(8 – 12) × (–11)

# Identity Element

When you consider the word "identity" in everyday language, you might think about the unique characteristics and traits that make a person distinct. In this sense, you can identify every number by its place on the number line. Even if a number has two different ways of presenting itself, like $\frac{1}{2}$ and 0.5, its position on the number line is fixed and unchanging. When we take a number and add something to it or multiply it by something else, we almost always convert it into a totally different number. There are two special cases in which the number is unchanged. These special cases are called *identity elements* because they allow the number to keep its identity.

When you're working with addition (or subtraction), the identity element is the number zero. That's all there is to it. When you add zero to a number it's unchanged. Subtracting zero from a number doesn't do anything. Working with zero is really, really easy. Kind of like watching TV, you don't really have to do anything.

**Example:** Calculate 87237546349 + 0.

**Solution:** 87237546349.

What about multiplication and division? When you're multiplying and dividing, the identity element is 1. Multiply by 1 or divide by 1? Easy: you get the original number back.

**Example:** Calculate 87237546349 × 1.

**Solution:** 87237546349.

When you're working with the identity element, even really big numbers are still easy.

You probably noticed a long time ago how easy it is to add 0 to a number or to multiply a number by 1. You might be wondering why we even bring it up now. You may even be starting to worry that something besides the sky is up. After all that trouble with the Distributive Law, could things really be this easy?

Well, yes and no. Yes, it is easy to add 0 or to multiply by 1; what makes it tricky is recognizing why you would use these ideas in a more complicated problem. At this point, you probably don't have much of a reason to do that, but there will be problems you'll face later where this is important.

If math were a movie, this would be the point where we'd cue the music that tells you that this seemingly innocent idea may come back in unexpected and important ways. Please play such music right … now!

**Timely Tips** _____

In some problems, adding zero or multiplying by one is quite useful! An example from fractions: suppose you wanted to know what ½ + ⅙ is equal to. First, we're going to multiply ½ by ³⁄₃. This does not change the value of ½, as we show you next.

3 ÷ 3 = 1, so ³⁄₃ also must equal 1, since fractions are also division problems.

Thus, ½ × ³⁄₃ = ³⁄₆. In other words, ½ and ³⁄₆ are the same value (you could find them at the same place on the number line).

And, finally, ³⁄₆ + ⅙ is just like taking three pieces of a pizza cut into six slices and then taking another slice! After we've multiplied by ³⁄₃ (that is, multiplied by 1), it becomes easy to find the answer: ⁴⁄₆.

# Inverse Operations

Have you ever known a couple who always bickers? If he says "up," she says "down." If he says "yes," she says "no." No matter what one says, the other says the opposite. The mathematical version of this would be *inverse operations*. Whatever one operation does, its inverse operation undoes.

Imagine you had a number that you liked and then someone went and added 4 to it. How would you undo this? How would you restore your original number? If you said that you'd subtract 4 back from it, you'd be exactly right. The way to undo adding 4 would be to subtract 4. This works for any number, not just 4, and it also works if you first subtract and then add. Because adding and subtracting undo each other, they are inverse operations.

**Example:** Calculate 23476 + 392847 – 392847.

**Solution:** If you had a calculator handy, you might be tempted to solve this problem by just typing in the numbers. But there's an easier way that makes use of inverse operations.

Because adding 392847 and subtracting 392847 undo each other, you are left with the original number 23476.

The inverse operation of multiplication is division. If you take a number and multiply by 7 and then divide by 7, you'll end up with the number that you started with.

**Example:** Calculate 329375 × 23 ÷ 23.

**Solution:** Because of inverse operations, dividing by 23 undoes multiplying by 23, so you get 329375.

You might wonder why anyone would ever ask you to do such an easy problem. In algebra you will be faced with complicated expressions in which several operations have been done to a number. You will use inverse operations to

**Practice Makes Perfect**

Problem 9: Calculate: 45353 + 45 – 45.

Problem 10: Calculate: 672834 × 253 ÷ 253.

Problem 11: Calculate: 9834 + (234 – 37) + (37 – 234).

Problem 12: Calculate: –23 × (56–21) ÷ 23 + 15 + 56 – 21.

undo all of those steps and get back the original number. It is kind of like using the undo command when one of your pets decides to type a few paragraphs of an e-mail for you.

## The Least You Need to Know:

- ◆ Evaluate statements in parentheses first.

- ◆ Do multiplications and divisions before additions and subtractions, working left to right except where you can apply the Associative or Commutative Law.

- ◆ Order and grouping don't matter when you only add or only multiply.

- ◆ In a problem with both adding and multiplying, you may have to use the Distributive Law.

- ◆ Addition and subtraction undo each other.

- ◆ Multiplication and division undo each other.

# Part 2

# Let's All Remain Rational

The question that is asked more than any other in math class is, "Will this be on the test?" If you're asking it about fractions, I can assure you that it will be. Fractions are on this test. They're probably on the next test. They'll be on most of the tests that you take in algebra class, and they come back in algebra 2, advanced math, pre-calc, and calculus.

You are going to be spending a lot of time with fractions if you take any more math. The reason that fractions keep coming back is not some sort of mathematical sadism or an attempt by math teachers to try to trick you with difficult things. It's because they are so useful. Every time you want to work with a whole that has been divided into parts, you'll use fractions.

*"So what's it going to be? Want 1/3rd of the cake or 3/8ths?"*

**4**

# Getting to Know Fractions

## In This Chapter

- ◆ Rewriting fractions in different forms
- ◆ Determining whether two fractions are the same
- ◆ Multiplying and dividing fractions
- ◆ Adding and subtracting fractions

You, your family, and your hunter-gatherer friends have finished consuming the wildebeest from Chapter 2. Your ability to divide the spoils of hunting into equal portions has attracted the attention of the clan's leader. He wants you to be the Chief Mathematician of the United Wildebeest Company, a major interclan corporation in charge of collection, division, and distribution of the majority of the wildebeest trade throughout the Western Valley. You certainly could use the massive allotment of wildebeests you are promised, but you've got to figure out how to deal with all the fractions you're going to run into at your new job.

You've spent years learning how to add, subtract, multiply, and divide. You've probably gotten plenty of practice doing those four operations with whole numbers and maybe some working with decimals. Now is the time for you to put those skills to good use as you learn about the arithmetic of fractions for your new job.

There's nothing inherently difficult about fractions other than there are a lot of things to keep in mind at once. You need to be able to flexibly think about a fraction either as a whole divided into parts or as a division problem waiting to be answered. You also need to be able to factor integers, carry out the four basic operations, and respect the order of operations. Not much at all. And we wonder why so many students (and teachers) struggle to understand fractions!

# Talking Rationally About Fractions

There's a lot of vocabulary associated with fractions. Beyond names for the parts of a fraction, such as its numerator and denominator, we also have names for different types of fractions and ways to express them.

One group of fractions that has a special name is the *improper fraction*. You might wonder what this group is up to. Have they been hanging out with the wrong crowd? Using the wrong forks to eat their salads? Saying things that should best be left unsaid? The only faux pas that they've committed is to have their numerators larger than their denominators. So $^{21}/_8$ would be an improper fraction. What's wrong with them to make them be called improper? Nothing. They are perfectly good fractions, and you will use them all the time. Stamp out ignorance about improper fractions and accept them as worthy numbers to use in your math problems.

*$^{21}/_8$ expressed in chocolate. Mmm ... Chocolate ....*

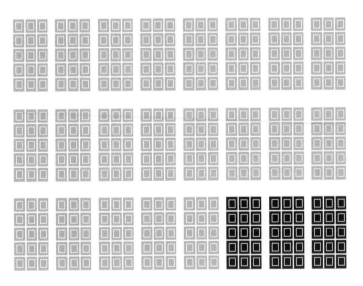

The only downside to improper fractions is they don't always give a good sense of the size of the number. If you have $^{21}/_8$ of a chocolate bar, it means that you've been breaking chocolate bars into eight evenly sized pieces each, and you've taken 21 pieces. It might be more descriptive to represent this as $2^5/_8$ bars of chocolate. Numbers that are written as an integer plus a fraction are called *mixed numbers*, and there's an invisible plus sign between the number and the fraction.

If you're going to be adding, subtracting, multiplying, or dividing fractions, you'll want to write mixed numbers as improper fractions. If you're trying to explain your answer to someone who isn't a math geek, you'll want to express yourself using mixed numbers instead of improper fractions. You've probably noticed that recipes are more likely to call for $4^1/_4$ cups of flour than for $^{17}/_4$ cups of flour.

To convert a mixed number to an improper fraction, you need to find both the numerator and the denominator. The denominator is easy to find: it's the same as the denominator of the fraction part of the mixed number. To find the numerator of the improper fraction, you take the whole number part of the mixed number, multiply it by the denominator of the fraction part, then add in the numerator of the fraction part.

**Example:** Convert $5^3/_{11}$ to an improper fraction.

**Solution:** The denominator will be 11 since the denominator of the fraction part of the mixed number is 11. To compute the numerator, you calculate $5 \times 11 + 3 = 58$. So as an improper fraction, this is $^{58}/_{11}$.

When converting back from improper fractions to mixed numbers, it's again easy to find the denominator. The denominator for the fraction part of the mixed number is the same as the denominator of the improper fraction. You can do the rest of the problem with division.

Remember that a fraction can be thought of as a division problem that has been asked but not yet solved; converting from an improper fraction to a mixed number gets you closer to the solution of that division problem. You will need to take your improper fraction and divide the numerator by the denominator. The quotient will give you the integer part of the mixed number, and the remainder will be the numerator of the fraction part.

| **Practice Makes Perfect** |
| --- |
| Problem 1: For each of the following fractions, determine whether it is an improper fraction or a mixed number, and rewrite it in a different form. |

(a) $^{23}/_5$

(b) $-5^5/_7$

**Example:** Convert $^{17}/_3$ into a mixed number.

**Solution:** You know that you'll have some integer and some thirds. Since the denominator of the improper fraction is 3, the denominator of the fraction part of the mixed number will also be 3. When we calculate $17 \div 3$, we get 5 with a remainder of 2. Therefore the mixed number is $5^2/_3$.

# Secret Agent Fraction

Not only does the collection of rational numbers sometimes improperly go by the other name "fractions," but individual fractions have multiple identities, too, just like secret agents and characters in a Russian novel. Though there are multiple ways to write these numbers, beneath all the spy gear each number has a specific value that is defined by its place on the number line. This is the fingerprint you can blow the number's cover with.

## Equivalent Fractions

Fractions don't take on aliases because they're trying to hide from their enemies. Or at least that's what they want us to think. One reason for the multiple representations of a fraction is because you can interpret a fraction as giving the relationship between parts and a whole. Whether you cut a cake into 12 equal pieces and take three of them or whether you cut it into 16 equal pieces and take four, you still have taken a quarter of the cake. (Did you leave enough for everyone else to have some of the cake?) Because they represent the same amount of cake, the fractions $^3/_{12}$ and $^4/_{16}$ are equal and both have the same value as $^1/_4$. We say that two fractions are *equivalent* if they both have the same value.

*The fractions two-eighths and four-sixteenths are equivalent.*

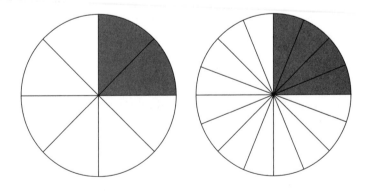

You don't want to have to bake cakes and then cut them up into pieces in order to tell whether two fractions are equivalent, and, fortunately, there are easier ways to tell. One method uses a process called cross-multiplying. This is the simplest and most effective interrogation technique for fractions. If two fractions are equivalent, when you take the numerator of the first fraction and multiply that by the denominator of the

**def•i•ni•tion** _____

Two fractions are **equivalent** if they both represent the same number. Equivalent fractions represent the same point on the number line. The easiest way to determine if they have the same value is to cross-multiply.

second, you'll get the same answer as when you multiply the numerator of the second by the denominator of the first.

**Example:** Use cross-multiplying to determine whether ⅔ and ⁶/₉ are equivalent.

**Solution:** Start by looking at $\frac{2}{3}$ and $\frac{6}{9}$ and multiplying $2 \times 9 = 18$. Next, you look at $\frac{2}{3}$ and $\frac{6}{9}$ and multiply $6 \times 3 = 18$. Both come out to the same number, 18, so the fractions are equivalent.

**Example:** Use cross-multiplying to determine whether ¾ and ¹⁰/₁₅ are equivalent.

**Solution:** Start with $\frac{3}{4}$ and $\frac{10}{15}$, multiplying $3 \times 15 = 45$. Next do $\frac{3}{4}$ and $\frac{10}{15}$, multiplying $10 \times 4 = 40$. Since 45 and 40 are different numbers, the fractions are not equivalent.

Another way to find equivalent fractions is to take a fraction and multiply both the numerator and the denominator by the same number. The new fraction is equivalent to the old one.

**Example:** Multiply both the numerator and the denominator of ²/₇ by 3 to get an equivalent fraction.

**Solution:** $2 \times 3 = 6$ and $7 \times 3 = 21$, so we get ⁶/₂₁. This fraction is equivalent to ²/₇.

**Example:** Find two different fractions that are equivalent to ⁵/₁₁.

**Timely Tips** _____

Cross-multiplication actually works by exploiting the idea of equivalent fractions. Do you see how? We'll go over an explanation of why cross-multiplication works in Chapter 12 when we have variables.

**Solution:** Unlike most math problems you might be used to seeing, this one does not have one right answer. In fact, there are infinitely many right answers and you just have to find two of them. If you're tempted to guess and hope that you'll find one of the infinite number of right answers, be careful because there are also an infinite number of wrong answers to this problem! Isn't that fascinating and scary?

---

**Practice Makes Perfect**

Problem 2: For each pair of fractions, use cross-multiplying to determine whether or not they are equivalent.

(a) ¾ and $^{12}/_{14}$

(b) $^5/_7$ and $-^{10}/_{14}$

(c) $^0/_2$ and $^0/_{100}$

---

For an open-ended problem like this one, you'll need to pick two numbers to multiply your numerator and denominator by. For the sake of this problem, you can pick 3 and 10, but any two numbers except 0 would work.

Since $5 \times 3 = 15$ and $11 \times 3 = 33$, one of your equivalent fractions is $^{15}/_{33}$. Because $5 \times 10 = 50$ and $11 \times 10 = 110$, another one is $^{50}/_{110}$.

This process also works if you divide both the numerator and the denominator of the fraction by the same number. This is the primary skill used when reducing a fraction to simplest terms.

## Reducing to Simplest Terms

Because every fraction has infinitely many equivalent forms, it can get confusing working with them. Just like how people with different nicknames tend to have just one legal name, the same thing happens with fractions. In order to clear things up a bit, we reduce the fraction to what's called *simplest terms* or *lowest terms*.

## def•i•ni•tion

A fraction is in **simplest terms** if its numerator and denominator have no common factors other than 1. Simplest terms is also called **lowest terms** because you can't make the denominator or numerator smaller integers and keep the same value for the fraction.

The process of reducing a fraction to lowest terms is based on the fact that dividing both the fraction's numerator and its denominator by the same number will give an equivalent fraction. It also relies on your ability to factor numbers into a product of primes. Don't worry about having to use two skills simultaneously to solve this type of problem. If you can walk and chew gum at the same time (or walk several steps, stop, chew a little, stop, walk several steps …) you already have the necessary multitasking ability. To brush up on your factoring skills, see Chapter 1.

**Example:** Reduce the fraction ⁶⁄₉ to simplest terms.

**Solution:** The numerator of the fraction can be factored as 6 = 2 × 3, and the denominator factors as 9 = 3 × 3. The fraction can be rewritten as $\dfrac{2 \times 3}{3 \times 3}$. There is a common factor of 3 in both the numerator and the denominator and it can be divided out of both. We write this as $\dfrac{2 \times \cancel{3}}{3 \times \cancel{3}}$, showing that you've divided out by the common factor of 3. Crossing-off factors usually works only if there is no addition or subtraction in either the numerator or the denominator. This leaves you with the fraction ²⁄₃.

**Example:** Reduce the fraction ⁴⁄₁₂ to simplest terms.

**Solution:** Factoring, 4 = 2 × 2 and 12 = 2 × 2 × 3. Dividing out both factors of 2 in the numerator and the denominator, you get $\dfrac{\cancel{2} \times \cancel{2}}{\cancel{2} \times \cancel{2} \times 3}$. In the numerator, you have calculated 4 ÷ 4 = 1, and in the denominator, you've calculated 12 ÷ 4 = 3. This leaves us with the fraction ¹⁄₃ in simplest terms.

## Kositsky's Cautions

If you divide out all the factors that appear in either the numerator or the denominator, then you are left with 1.

Don't forget that you can never divide a nonzero number by anything and get zero! For example, $\dfrac{2}{6} = \dfrac{\cancel{2}}{\cancel{2} \times 3} = \dfrac{1}{3}$. After canceling the two, the numerator is one!

## Practice Makes Perfect

Problem 3: Reduce the following fractions to lowest terms.

(a) ⁶⁄₉

(b) ⁴⁄₈

(c) ¹⁵⁄₉

# Fractions and Decimals

It can be pretty easy to turn a fraction into a decimal—especially if you have a calculator handy. Fractions can be thought of as unfinished division problems; to convert them into decimals, you just take the numerator and divide it by the denominator. Both the original fraction and the newly found decimal are the exact same number, just written in two different ways, and both occupy the exact same point on the number line. Another alias uncovered!

> **Kositsky's Cautions** _____
>
> While it is true that you can write any number as a fraction by putting it over 1, this, generally, is not what is meant when we convert decimals to fractions. Trying to convert $\pi$ to a fraction by writing it as $\frac{\pi}{1}$ is cheating. When possible, we want the fraction to be a rational number—a fraction where both the numerator and the denominator are integers.

Turning decimals into fractions is a bit trickier, however, and it doesn't work with all decimals. Not every decimal can be turned into a fraction. This is a case where the problem is not just hard, it's downright impossible. The only decimals that can be written as fractions are rational numbers. You can recognize a rational number's decimal expansion because it either terminates or it repeats.

## Terminating Decimals

As sinister as a terminating decimal may sound, it's one of the nicer types of decimals to work with. There will be some numbers after the decimal point and then it will stop. This includes pleasant decimals like 0.2 and 18.53 as well as somewhat more involved ones like 0.28954626348349. (That's a period, not a second decimal point.)

> **Kositsky's Cautions** _____
>
> You can't trust a calculator to tell you whether a decimal repeats or not. Some decimals have repeating patterns so long they don't fit on your calculator screen. For example, $\frac{1}{17}$ doesn't repeat for 16 digits!

Converting this type of decimal into a fraction is relatively straightforward. The numerator is the integer that you get when you ignore the decimal point and use all the digits. The denominator is found by counting how many numbers are to the *right* of the decimal point and taking 1 and following it by that many zeros.

**Example:** Convert 0.63 to a fraction.

**Solution:** This is a terminating decimal because its decimal expansion just stops.

As a fraction, the numerator would be 63 because you ignore the decimal point to find the numerator.

The denominator is found by counting the number of digits to the right of the decimal point, in this case two, and taking a 1 followed by that many zeros. This gives you a denominator of 100 (two zeros, one for each of the numbers to the right of the decimal) so your fraction is $^{63}/_{100}$.

**Example:** Convert 1.23487463 to a proper fraction or mixed number.

**Solution:** Another terminating decimal. This one is longer and has an interesting twist. It's greater than 1!

As a fraction, the numerator would be 123487463 since we ignore the decimal point.

The denominator is once again found by counting the number of digits to the right of the decimal point, in this case eight, and taking a 1 followed by that many zeros. This gives you a denominator of 100,000,000.

Thus, your fraction is $\dfrac{123,487,463}{100,000,000}$. However, this is an improper fraction. Though I don't want to discriminate against them, the problem asks for proper fractions or mixed numbers. Math is not a good time for civil disobedience.

Changing this fraction into a mixed number is particularly simple because the denominator of 100,000,000 is a *power of ten*, a one followed (or preceded) by some number of zeros. We subtract as many hundred-millions as possible from the numerator.

$$\frac{123,487,463 - 100,000,000}{100,000,000} = \frac{23,487,463}{100,000,000}$$

Then, we put the number of hundred-millions we subtracted to the left of the fraction to make a mixed number, our final answer $1\dfrac{23,487,463}{100,000,000}$.

## Repeating Decimals

Do you know people who just go on and on about the same thing when they talk? Same thing over and over, again and again? That's what it's like with *repeating decimals*. You might have seen 0.333333 …, the most common example of a repeating decimal. Any decimal that eventually starts repeating the same exact fixed pattern is a repeating decimal. This would include decimals like 37.0808080808 … and ones like 2347.27549557557557557 …. We'll also write these as $37.\overline{08}$ and $2347.27549\overline{557}$ with the overbar meaning "repeating the sequence underneath me forever and anon." The overbars are generally a bit melodramatic, but they probably just feel bad that they can't go to the left of the decimal point.

> **Practice Makes Perfect**
>
> Problem 4: Convert each of the following decimals into a fraction.
>   (a) 0.47
>   (b) 0.5684379

**Practice Makes Perfect**

Problem 5: Convert each of the following repeating decimals into a fraction.

(a) $0.\overline{1}$

(b) $0.\overline{84}$

Thankfully, it is possible to turn any of these long-winded decimals back into a fraction, but the arithmetic sometimes gets pretty tricky and involved. For now let's stick with decimals that are between 0 and 1 and where every digit is part of the repeating pattern, like $0.\overline{345}$ or $0.\overline{5}$. In that case, all you do is count how many digits long the repeating part is, and you write the repeating part over that many nines. So $0.\overline{345}$ would be $^{345}/_{999}$ and $0.\overline{5}$ would be $^5/_9$.

This method is a shortcut, but it only works for those special types of repeating decimals. If there is even one number present that is not part of the repeating pattern, this method will not work. It's great for numbers like $0.\overline{7}$, but it doesn't work at all for numbers like $0.8\overline{79}$. The 79 keeps repeating, but there's only one eight. To convert this slightly trickier decimal to a fraction, you need to use a different method that either relies on algebra or gets very complicated.

**Timely Tips** _____

If a decimal repeats, you can choose how many copies of the repeating part to write. Putting more or less of it on your paper doesn't change the value of the decimal as long as the overbar or ellipsis is used properly.

# Multiplying and Dividing Fractions

Compared to adding and subtracting fractions, multiplying and dividing them is very easy. In fact, multiplying fractions is one of the easiest things you could possibly do with fractions, except possibly looking at them or tossing them in the trash. All you have to do is multiply straight across the numerators and the denominators. That's it!

**Example:** Calculate $\dfrac{2}{3} \times \dfrac{4}{9}$.

**Solution:** $\dfrac{2}{3} \times \dfrac{4}{9} = \dfrac{2 \times 4}{3 \times 9} = \dfrac{8}{27}$.

One of the best tips in math is that when you find an easy way to solve a problem, you want to be sure to apply it to any situation where it will help you. We're going to use the simplicity of multiplying fractions to our advantage and turn every fraction division problem into a multiplication problem. Because multiplication and division are not the same thing, we need to introduce a slight twist—or, rather, a flip—to convert a multiplication problem into a division problem.

Let's first do a special kind of multiplication of fractions, multiplying by the *reciprocal*. To form the reciprocal of a fraction you switch the numerator and the denominator. For example, the reciprocal of ³/₇ is ⁷/₃. In a sense, you're flipping the fraction on its head. Don't worry, this doesn't hurt even the most fragile fractions. Think back to Chapter 3 where we discussed inverse operations. If we start with some quantity, divide by a number, and then multiply by the same number, we miraculously get back our original quantity. Just as this is true for all real numbers, it is true for fractions. Thus, ⁵/₃ ÷ ⁵/₃ = 1.

From what you just learned about fraction multiplication, you can see that $\frac{5}{3} \times \frac{3}{5} = \frac{5 \times 3}{3 \times 5} = \frac{15}{15} = 1$. Notice anything strange? It looks like multiplying by ³/₅ does the same thing as dividing by ⁵/₃! In fact, this is true for all fractions; to divide fractions, you multiply by the reciprocal of the second number. Let's look at a quick example.

**Example:** Calculate ⁶/₁₇ ÷ ⁴/₅.

**Solution:** Convert this to a multiplication problem and multiply by the reciprocal, getting ⁶/₁₇ × ⁵/₄, which is $\frac{6 \times 5}{17 \times 4} = \frac{30}{68}$ . Simplifying, the answer is ¹⁵/₃₄.

Because a fraction is a division problem, division of fractions sometimes can be written as a fraction over another fraction, which is known as a *complex fraction*, like $\frac{\frac{2}{3}}{\frac{4}{9}}$ . This is the same as ²/₃ ÷ ⁴/₉ = ³/₂. Whenever you're faced with a complex fraction, you'll usually want to rewrite it as a division problem.

**Practice Makes Perfect**

Problem 6:  Calculate $\frac{9}{7} \times \frac{2}{5}$ .

Problem 7:  Calculate $\frac{2}{5} \div \frac{9}{7}$ .

**def•i•ni•tion**

A **complex fraction** is one where the numerator and denominator are both fractions.

# Adding and Subtracting Fractions

If I had a nickel for every time one of my college students made a mistake adding fractions, I would be able to buy a very fancy cup of coffee at one of those overpriced coffee shops on a fairly regular basis. This is one of the times that you must know the rules and follow them slavishly. If there was ever a good time to be a math zombie and to act without thinking, now would be that time.

**Kositsky's Cautions**

Since fractions are secretly division problems in disguise, you need to keep in mind the order of operations when working with complicated fraction problems like $\frac{4+7}{13+9} + \frac{-6(-7)}{2\times 2}$. You need to imagine there is a set of parentheses around the entire numerator and a set of parentheses around the entire denominator. In some problems it helps to interpret the fraction bar as a division sign.

## Fractions with Like Denominators

Some fraction problems are fairly easy. Imagine that we have baked another cake and cut it into 16 pieces. If you take 3 pieces, you'll have $3/16$ of the cake. What if you take 4 more pieces of cake? This would be represented mathematically by $3/16 + 4/16$. Your common sense should tell you that if you have 3 pieces and then take 4 more, then you should now have 7. Because your cake is still cut into 16 pieces, you represent the amount of cake you have as $7/16$.

This works because our cake is cut into equal pieces so we're dealing with pieces that are all the same size. Mathematically speaking, we describe this situation by saying that we're adding or subtracting fractions with the same denominator. When you add, you get $3/16 + 4/16 = 7/16$. When you are adding (or subtracting) fractions that have the same denominator, you add (or subtract) the numerators while keeping the denominator the same.

### Practice Makes Perfect

Problem 8: Calculate the following expressions.

(a) $1/13 + 9/13$

(b) $5/8 − 9/8$

**Example:** Calculate: $7/11 + 2/11$.

**Solution:** Since the fractions have the same denominator, we add the numerators, resulting in $9/11$.

**Example:** Calculate: $8/13 − 5/13$.

**Solution:** Again, the fractions have the same denominator, so we can subtract the numerators, getting $3/13$.

# Fractions with Unlike Denominators

Maybe you want to add together two fractions like $\frac{1}{2}$ and $\frac{1}{3}$. Think about what this means in terms of the cake. If you have half the cake and then take another third of the cake, what fraction of the cake will you have now? If you know anything about cake, you should expect to have a lot of it. Almost all of the cake, even. You certainly won't have less than half the cake if you start out with half and then take even more! If math is going to make any sense at all, we need to have $\frac{1}{2} + \frac{1}{3}$ be a fraction that is bigger than $\frac{1}{2}$ and smaller than 1.

As long as your fractions have different denominators, you can't do anything about adding or subtracting them. Your first step is to convert them into equivalent fractions that have the same denominator. To give two fractions the same denominator, we say that we put them over a common denominator. We usually try to put them over the *lowest common denominator*. When people talk about the lowest common denominator in terms of culture and society, it's usually a bad thing. In math the lowest common denominator is a good thing. Your life is much easier when working with fractions once they share a common denominator—lowest or otherwise.

Simply speaking, the lowest common denominator is the least common multiple of the two denominators. If your fractions are $\frac{3}{14}$ and $\frac{4}{21}$, then your lowest common denominator would be the least common multiple of 14 and 21, which is 42. Finding the lowest common denominator is fairly easy. Rewriting part of the fractions so that they share this denominator is easy (just write $\frac{?}{42}$!) The difficulty comes in correctly rewriting the numerators.

**Kositsky's Cautions**

You might be tempted to come up with your own method for adding fractions. The obvious shortcut that comes to mind is completely wrong and should never be used. You know what I'm thinking. Don't do it!

# def•i•ni•tion

The **lowest common denominator** is the smallest integer such that both denominators divide evenly into that integer. The lowest common denominator is always the least common multiple of the denominators.

**Example:** Rewrite $\frac{1}{2}$ and $\frac{1}{3}$ as equivalent fractions that share a common denominator.

**Solution:** The lowest common denominator is the least common multiple of 2 and 3, which is 6. You need to rewrite both of these fractions individually as equivalent fractions with a denominator of 6.

You'll be converting each fraction to an equivalent fraction by multiplying both its numerator and denominator by the same number.

**Timely Tips** _____

The process of creating equivalent fractions is simply multiplying a fraction by a cleverly written 1. Because 1 is the multiplicative identity element, multiplying by 1 doesn't change the value of your fraction.

To convert $\frac{1}{2}$ to an equivalent fraction with a denominator of 6, first forget that $\frac{1}{3}$ is in this problem and that you're adding fractions at all! If trying to forget about adding fractions is making you sad, you're probably destined to be a math major.

Right now the only thing in your mind should be a problem about equivalent fractions: convert $\frac{1}{2}$ to an equivalent fraction with a denominator of 6.

Then you need to figure out what you multiply 2 by in order to get 6. Divide 6 by 2 and get 3.

Next you multiply both the numerator and the denominator by 3 to get the equivalent fraction $\frac{3}{6}$.

Dredge through the depths of your short-term memory for whatever it was you were trying to forget about just now. Conveniently remember that $\frac{1}{3}$ needs to be converted into an equivalent fraction and forget everything else. This may seem very difficult. However, if you work at it long enough, one day your mind may be so highly trained that not only do you forget what you want to forget but you also forget what you want to remember!

We're going to apply similar steps to $\frac{1}{3}$. In this case, to get the denominator to be 6, you have to multiply both the numerator and the denominator by 2, like $\frac{1\times2}{3\times2}$, giving you an equivalent fraction of $\frac{2}{6}$. Therefore, when rewritten as equivalent fractions with a common denominator, $\frac{1}{2}$ and $\frac{1}{3}$ are $\frac{3}{6}$ and $\frac{2}{6}$.

**Example:** Rewrite $\frac{5}{14}$ and $\frac{4}{21}$ as equivalent fractions with a common denominator.

**Solution:** The least common multiple of 14 and 21 is 42, so the lowest common denominator of the two fractions is 42. In order to get $\frac{5}{14}$ over that denominator, you have to multiply both the numerator and the denominator of $\frac{5}{14}$ by 3 (because $14 \times 3 = 42$), resulting in $\frac{15}{42}$. To get $\frac{4}{21}$ to be over 42, you multiply both the numerator and the denominator of $\frac{4}{21}$ by 2 (because $21 \times 2 = 42$), yielding $\frac{8}{42}$. Rewritten with a common denominator, the fractions are $\frac{15}{42}$ and $\frac{8}{42}$.

**Timely Tips**

Any common denominator will work. One shortcut is to multiply the denominators together. However this method can cause you to work with some very large numbers. For example, in the previous example multiplying together the denominators would make us calculate $21 \times 14$ and then have to work with 294 as our denominator!

It's traditional and often faster to use the lowest common denominator. Most important, if you want to be cool like all the mathematicians, you should use the lowest common denominator, too.

The process of getting fractions over a common denominator makes it possible to add or subtract any two fractions. Just get them both over a common denominator and then add.

**Example:** Calculate: $\frac{3}{14} + \frac{4}{21}$.

**Solution:** Getting those fractions over a common denominator allows you to rewrite the problem as $\frac{9}{42} + \frac{8}{42}$. Once over a common denominator, this is easy as pie (or cake). Just add the numerators and keep the denominator the same, giving you $\frac{17}{42}$. Isn't that cool?

| **Practice Makes Perfect** |
| --- |
| Problem 9: Calculate $\frac{1}{2} + \frac{1}{5}$. |
| Problem 10: Calculate $\frac{2}{3} + \frac{7}{6}$. |
| Problem 11: Calculate $\frac{5}{6} - \frac{9}{8}$. |

## The Least You Need to Know

- The same number can be represented as a fraction in many different ways.
- You can only add or subtract fractions when they have common denominators.
- To add or subtract fractions without common denominators, you must convert them both to equivalent fractions with common denominators.
- To multiply fractions, multiply the numerators then multiply the denominators.
- To divide fractions, multiply by the reciprocal of the second fraction.

# Chapter 5

# Tips on Percentages

## In This Chapter

- ◆ The meaning of percents
- ◆ Converting between percents, fractions, and decimals
- ◆ Calculating with percents
- ◆ Applying percents to finance and science

Percents live in a sort of middle ground between fractions and decimals. They function as fractions whose denominator is 100. Because 100 is a power of 10 (and our number system is base 10), they also can be easily converted into decimals. The "cent" part of "percent" means "hundred" in Latin, so "percent" can be thought of as meaning "per hundred."

Most applications of percents fall into one of three main categories: fractions, rates, or comparisons. If you heard a rumor that 95 percent of the people buying this book believed it improved their lives, then the percent would be describing a fraction of the people. The percent would be a rate if it described your happiness as improving by 10 percent. To use a percent as a comparison, you might say that people who love math are 150 percent happier with their lives than people who don't.

There are a wide variety of word problems that arise from percents. Some of them come from science and involve concentrations of chemicals and mixture problems. Many of them come from the world of money and finance. Despite any claims you might hear to the contrary, the most common application of percents in your math class probably will be in determining your grade. Even if you fall short of earning 100 percent in math class, once you know about percents, you'll be able to calculate the scores you need to earn that A.

# What Are Percents?

Percents are just another way to look at fractions, but this time all the fractions have the same denominator: 100. When working with problems involving percents, it's essential to be able to identify both the whole object as well as the part being considered. When you're looking at a score on a math test, the whole represents all the points available and the part shows how many of them you earned. In a problem about the percent of ethanol in an aqueous solution, the solution is the whole, and the ethanol is the part.

### Kositsky's Cautions

Don't get all excited upon hearing that percents already have a common denominator. In many problems involving percents, trying to add or subtract the percents will give you the wrong answer.

When you're working with percents, 100 percent represents the entire whole. A score of 100 percent means that you earned all the points on the exam. If 100 percent of the jellybeans are cherry, then they're all the same flavor. Depending on the circumstances, it may or may not make sense to have a percent that is greater than 100 percent. If the percent is representing a fraction and it doesn't make sense to have more than the whole, then you can't have more than 100 percent. If there are 100 students in a class, you can't have more than 100 of them named David. However, if you're comparing two different collections and one is bigger than the other, you might want to talk about one being more than 100 percent of the other. If I have 100 jellybeans and you have 149, then you will have 149 percent of what I have.

## Percents and Decimals

Every fraction can be written as a decimal, including percents. To convert a percent into a decimal, you simply move the decimal point two places to the left. If you're looking at a percent that is less than 10 percent, you will have to add zeros to the beginning of the percent so that you have enough room to place the decimal point without violating the rules of mathematics.

**Examples:** Convert the following percents into decimals:

(a)     17%

(b)     6%

(c)     4.2%

(d)     0.01%

(e)     156%

**Solutions:**

(a)     17%̸       Get rid of the percentage sign and add the implied decimal point.

         17.        Move the decimal point over two places to the left.

         0.17       We discover our final answer of 0.17

(b)     6%̸        Get rid of the percentage sign and add the implied decimal point.

         6.         Move the decimal point over two places to the left.

         0.06       We discover our final answer of 0.06

(c)     4.2%̸       Get rid of the percentage sign.

         4.2        Move the decimal point over two to the left.

         0.042      We discover our final answer of 0.042

(d)     0.01%̸      Get rid of the percentage sign.

         0.01       Move the decimal point over two to the left.

         0.0001     We discover our final answer of 0.0001

(e)     156%̸       Get rid of the percentage sign and add the implied decimal point.

         156.       Move the decimal point over two to the left.

         1.56       We discover our final answer of 1.56

Converting back from a decimal to a percent is easy: you just move the decimal point two places to the right.

**Examples:** Convert the following decimals into percents:

(a)      0.73

(b)      0.056

**Solutions:**

(a)      0.73                   Move the decimal point over two places to the right.

          73                     Add a percentage sign.

          73%                    We discover our final answer of 73%

(b)      0.056                  Move the decimal point over two places to the right.

          5.6                    Add a percentage sign.

          5.6%                   We discover our final answer of 5.6%

---

### Practice Makes Perfect

Problem 1: Convert the following percents into decimals.
   (a) 98%
   (b) 0.005%
   (c) 4.23%
   (d) 623%
Problem 2: Convert the following decimals into percents.
   (a) 0.37
   (b) 1.01
   (c) 3.3
   (d) 72

---

# Percents and Fractions

Since every fraction can be written as a decimal and every decimal can be written as a percent, it's not surprising that there's a link between fractions and percents. To convert a fraction into a percent, just convert the fraction to a decimal by dividing the numerator by the denominator and then convert the decimal into a percent.

**Examples:** Convert the following fractions to percents:

(a) $\frac{4}{5}$

(b) $\frac{17}{16}$

**Solutions:**

(a) $\frac{4}{5}$ = 0.80, which is 80%.

(b) $\frac{17}{16}$ = 1.0625, which is 106.25%

To convert a percent into a fraction, use the percent as the numerator and use 100 as the denominator. If you have a decimal in the numerator, multiply both the numerator and the denominator by the some power of 10 to move the decimal point all the way to the right. Once you have your percent as a fraction, simplify it to lowest terms.

**Examples:** Convert the following percents to fractions and reduce to lowest terms:

(a) 50%

(b) 2.4%

(c) 113%

**Solutions:**

(a) $\frac{50}{100} = \frac{1}{2}$

(b) $\frac{2.4}{100} = \frac{24}{1,000} = \frac{3}{125}$

(c) $\frac{113}{100}$

**Timely Tips**

If you're converting a fraction into a percent and the decimal expansion doesn't terminate, then you'll probably have to round. If the problem doesn't tell you how many decimal places to round to, then it's usually safe to round the percent to two decimal places. We would write $\frac{1}{7}$ = 0.14285714 ... as 14.29%.

---

**Practice Makes Perfect**

Problem 3: Convert the following fractions to percents.

    (a) ¾

    (b) $^{7}/_{20}$

    (c) $^{29}/_{25}$

    (d) 3⅖

Problem 4: Convert the following percents to fractions and reduce to lowest terms.

    (a) 25%

    (b) 9.8%

    (c) 0.04%

    (d) 355%

---

# Calculating with Percents

You'll need to keep in mind the dual nature of percents and their part-decimal, part-fraction existence if you want to calculate with them. If you're talking to Clark Kent, it would be worth your while to remember that he's also Superman. And if you're fighting with Superman, you should be aware you may be a front-page story in the news tomorrow—even if there are no eyewitnesses! You'll need to keep in mind everything that you learned about fractions to work properly with percents. When you're looking at two collections, you'll have to decide whether one of them represents a part of the whole or not. Remembering these two principles will help a lot when working with percents.

**Kositsky's Cautions**

When calculating with percents, always convert the percent into a decimal (or a fraction) before starting your computation. When working with 10%, you can use 0.10 or $^{1}/_{10}$, but never, never, never use 10.

## Finding Percentages

Here's the most important thing you need to know about these problems: when you're dealing with fractions or percents, the word "of" is a codeword telling you to multiply. If there are 28 bands playing at the outdoor music festival and 75% of them are terrible, then you can calculate the number of bands that are terrible by computing $28 \times 0.75 = 21$. It's probably not worth going unless the other 25% (7) of the bands are really good.

Turning this type of problem around, sometimes you're given the two quantities, and you're asked to find some percent-related relationship between them. The easiest ones are when you can quickly determine that one represents a whole and the other a part of the whole. In that case you calculate $\dfrac{\text{part}}{\text{whole}}$ and convert it to a percent.

**Example:** There are 512 chocolate candies in a bag, and 83 of them are red. What percent of the candies are red?

**Solution:** It's pretty clear that all the candy is the "whole" and the red ones represent a part of that whole. We calculate $83 \div 512 = 0.1621 = 16.21\%$.

What makes percent problems tricky at times is that it's not always clear what to consider as the part and what to consider as the whole. Suppose you have a dinosaur farm and you are raising two types of dinosaurs; you have 30 triceratops and 48 stegosauruses. If you wanted to know what percent of your dinosaurs are triceratops, then the 30 triceratops would be the part and all 78 of your dinosaurs (30 + 48) would be the whole. In that case, you would have $^{30}\!/_{78} = 0.3846 = 38.46\%$ of your dinosaurs being triceratops.

If, however, you were asked to compare the two different types of dinosaurs to each other (and not to the whole population) the calculation would be a bit different. In this case, the question probably would be somewhat awkwardly phrased (this isn't an idea that's expressed in English very often). You could say that you have $^{30}\!/_{48} = 0.625 = 62.5\%$ as many triceratops as you do stegosauruses. Or you could say that you have $^{48}\!/_{30} = 1.6 = 160\%$ as many stegosauruses as you do triceratops.

You need to be on the lookout for the language being used in percent problems. If you have 4 cookies and your brother has 125% *as many* cookies as you, then he has $4 \times 1.25 = 5$ cookies. However, if you have 4 cookies and he has 125% *more* cookies than you, then he has $4 + 4 \times 1.25 = 4 + 5 = 9$. When you see a problem with "percent as many as," then you just multiply by the percent (written as a decimal). If, however, the problem says "percent more than," you need to add the original amount plus the original amount times the percent (written as a decimal).

There's another way to think about situations that use the words "as many" with a percentage that's more than 100%. You can subtract 100 from the percentage and change the words to "more than." Having 125% as many cookies as your brother is the same as having 25% more than him. Having 175% as many goldfish as your sister is the same as having 75% more than she has.

Sometimes you may see a multi-step problem that combines these ideas.

**Example:** There are 250 boys at the school dance, and there are 10% more girls than boys at the dance. What percent of the students at the dance are girls?

**Solution:** What you're looking for in the end is the percent of girls at the dance, which you'll calculate as the number of girls divided by the total number of students at the dance. In order to find this, you first need to find the number of girls. You're given that there are 250 boys and that there are 10% more girls than boys, so you can calculate the number of girls as $250 + 250 \times 0.10 = 250 + 25 = 275$. Now you can finish the problem: there are 275 girls and $250 + 275 = 525$ students, so $^{275}/_{525} = 0.5238 = 52.38\%$ of the students at the dance are girls. Rather unfortunate for the girls.

---

### Practice Makes Perfect

Problem 5: There are 125 pens on the desk, and 120 of them are black. What percent of the pens are black?

Problem 6: A monster eats 160 math books for breakfast, and he reads 25% of the books before eating them. How many books does the monster read each morning?

Problem 7: There are 5 elephants in the zoo, and there are 20% more tigers than elephants in the zoo. How many tigers are there in the zoo?

---

## Percent Change

When you are looking at something changing, whether it's the price of a stock or your grade in a math class, you can use percents to express the change. In this type of problem, the whole will always be the initial value, and the part will be the new value minus the old value.

**Example:** Last year the crazy cat lady had 12 cats living in her house. This year she has 15 cats. What is the percent change in the number of cats living with the crazy cat lady?

**Solution:** The whole is the initial value, so the whole is 12. The part is the new value minus the old value, $15 - 12 = 3$. Thus, the percent change is $^{3}/_{12} = 0.25 = 25\%$, so there has been a 25% increase in the number of cats living in the house.

If the new value is less than the old value, then the new value–old value will be negative and you have a decrease.

**Practice Makes Perfect**

Problem 8: In 1956 the world's population was 2.8 billion. In 2006 it was 6.5 billion. What is the percent change in the world's population over the 50 years?

# Financial Examples

Here is your chance to practice what you've learned about percents in several common situations that come up in real life. Percents are a very common way that rates are expressed in situations dealing with banking and business.

## Interest

Here's your chance to get rich with math. What's the catch, you ask? Well, there are two. The first is that you need some money to get started with. The money that you start with is called the *principal*. The second is that you can't get rich quick—but you can get very rich slowly. Most people don't understand interest and its awesome power to create or destroy wealth.

Interest is quoted as a percentage, often as a yearly rate. A savings account that pays an interest rate of 4% means that every year you'll receive 4% of your money in interest. Similarly, a loan that charges 7% means that every year you'll pay 7% of the loan balance in interest. It should be pretty clear that you're better off being the one receiving interest than the one who is paying it. Keep that in mind, and you'll be well on your way to a rosy financial future.

Let's say you have $150 in a savings account that pays 4.2% interest and you want to know how much money you'll have in the account next year. The amount of interest that you earn is 4.2% of the *balance* in the account, so you'll make $150 × 0.042 = $6.30 in interest. This is added to your balance, so you'll have a total of $150 + $6.30 = $156.30 in your bank account.

**def•i•ni•tion**

When you make an initial deposit into a savings account (or borrow money for a loan), that amount of money is called the **principal**. The amount that you have in a savings account (or owe on a loan) at any given time is called the **balance**.

Where this gets complicated is figuring out what happens next year and the year after that and the year after that. If you take the $6.30 out of your bank account and go buy a sandwich, next year you'll get another $6.30. If you take the $6.30 and go and spend it, each year it will replenish itself. You won't make very much progress financially. Years down the road you'll still have your initial deposit of $150, and you will have eaten many fine sandwiches, but you won't be closer to any savings goals you have set.

Where the magic of mathematics can help you is if you leave your interest in the bank. This is called compound interest and is what happens if you earn interest on your interest. And then you can earn interest on the interest on your interest. Boring as it might sound, this is one of the surest ways to become wealthy.

**Example:** You put $150 in a bank account that pays 4.2% interest per year. If you allow the interest to accrue but don't withdraw any of it, how much money will you have after two years?

**Solution:** This requires some financial willpower as well as careful work with percentages. The first step is to figure out how much is in the account after one year, which you discovered before is $150 + $150 × 0.042 = $156.30.

Now you need to figure out what happens in the second year. Your new interest in the second year will be $156.30 × 0.042 = $6.56. Adding this to your balance, after the second year, you're up to $162.86.

While $162.86 doesn't sound like the sort of big bucks that will let you retire young and buy your own island in the Pacific, the same principle applies to your investments. If you invest money and let the interest build up over the long term, the power of the exponential function will reward you.

**Timely Tips**

Calculating compound interest efficiently requires the use of exponents, which are introduced in Chapter 7. You can do even more complicated financial calculations once you've mastered pre-calculus. Who said you couldn't get rich by doing math?

**Practice Makes Perfect**

Problem 9: Marvin has a savings account with 4.5% interest. If he deposits $2,500 into the account, how much money is in the account after four years?

**Example:** You deposit $1,200 into a savings account that pays 5.3% interest. If you leave the interest in the account, how much money is in the account after three years?

**Solution:** The interest you earn in the first year is $1,200 × 0.053 = $63.60, so after the first year you have $1,200 + $63.60 = $1,263.60 in the account.

In the second year, you'll earn $1,263.60 × 0.053 = $66.97, bringing your account up to $1,263.60 + $66.97 = $1,330.57.

In the third year, you'll earn $1330.57 × 0.053 = $70.52, bringing your total up to $1330.57 + $70.52 = $1,401.09.

# Sales Tax

When I was a little kid, we lived a few blocks from a small grocery store and I still remember the first time that I went on my own to buy candy. After I carefully made my choices and counted my money, I was shocked when the cashier told me that I owed a few more cents. That was my harsh introduction to the world of sales tax.

There are two aspects to sales tax. One is the amount of the tax; this is the amount that the government is interested in because it's the money that they get. The other is the total amount that you pay for the item plus the tax; you're more often interested in this part because this is what lets you decide if you have enough money to buy the item.

The tax rate is given as a percentage, such as 5%. This means that 5% of every purchase is collected as tax. Since this is a "percent of" problem, you'll be multiplying. You multiply the price by the tax rate (written as a decimal) to get the amount of tax. Then you add the tax to the price to get the price with tax since tax is an additional cost. Learning this simple principle will save you the heartache that I experienced.

**Example:** In a state with 6% sales tax, you buy a video game for $44.99. How much sales tax do you pay? What is the price with the tax?

**Solution:** 6% as a decimal is 0.06. To find the tax, you calculate $44.99 \times 0.06 = 2.6994$. Since this is a financial problem, you round to the nearest cent, so the tax will be $2.70. To determine the total amount that you pay to the store before you can bring home your game and start slaying vampires, you need to add the tax to the price, costing you $44.99 + $2.70 = $47.69.

**Timely Tips**

To calculate the total with the tax, you can multiply by 1 + tax rate. If you buy something for $25 and pay 7% tax, the total will be $25 \times 1.07$. This works because of the Distributive Law: $25 \times 1.07 = 25 \times (1 + 0.07) = 25 \times 1 + 25 \times 0.07 = 25 + 25 \times 0.07$.

# Tipping

You've taken your beloved to a fancy dinner at the Chateau Riche. The bill comes. Do you really want to get out one of those pre-printed tip charts from your wallet? Or are you going to impress your beloved by calculating the tip in your head?

The trick is that most common tip amounts are easy to compute if you can find 10%, and 10% is one of the easiest percentages to calculate in your head. To find 10%, all you need to do is take the total and then move the decimal point one place to the left. If you had a bill of $37.85, 10% of that would be $3.79 (when rounded to the nearest cent).

If you had a calculator with you, you'd calculate a 20% tip by taking the total and multiplying by 0.20. Here's where the laws of math can help you: multiplying by 0.20 is the same as first multiplying by 0.10 and then multiplying by 2.

**Example:** Find a 20% tip on a bill of $47.28 without a calculator.

**Solution:** 10% of $47.28 is $4.73. To find the tip you need to double $4.73. Because we're working in our heads and a tip can be off by a few cents, we can round $4.73 to a number that's easier to work with, like $4.70. Doubling, we get $9.40.

Another common tip amount is 15%. It's not as easy to calculate in your head as 20%, but it's still reasonable. In this case the laws of math come to our rescue again. To calculate 15%, you can first take 10% and then add half again as much.

**Example:** Find a 15% tip on a bill of $60.48 without using a calculator.

**Solution:** 10% of $60.48 is $6.05.

Rounding to an easy number to work with, we can use $6.00.

Half of $6.00 is $3.00.

Adding, we get $6.00 + $3.00 = $9.00, so our tip is $9.00.

> **Practice Makes Perfect**
>
> Problem 10: Find a 10% tip on a bill of $37.58 without a calculator.
>
> Problem 11: Find a 20% tip on a bill of $78.65 without a calculator.

So how much should you tip? Customs on tipping vary regionally and are thus almost certainly confusing. To make things worse, the preferred percentage and whether you tip on the total or on only part of the bill also depends on where you are. Aside from restaurants, tipping is common for many services. Which ones? I haven't a clue. Maybe this is why nobody gives me very good service anywhere. If in doubt, I'd suggest asking someone who works in the industry—usually not the person you could potentially be tipping.

## Commissions

A lot of people in sales work on commission, which means that they get a certain percentage of each sale that they make. The more they sell, the more money they take home. For most people, this is very strong motivation to work hard. I once heard a businessman say that he was looking for a way to put his receptionist on commission!

Knowing how commissions work could help you understand others' motivations and possibly provide a way for you to make enough money to go buy that Pacific island you've been eyeing.

Commission is calculated as a percentage of the selling price.

**Example:** A real estate agent earns a 6% commission on each house she sells. How much does she make if she sells a house for $248,000?

**Solution:** The commission is the percent of the selling price. $248,000 × 0.06 = $14,880.

# Profit

There are a few ways to express profit in terms of percentages. Unlike the previous problems you've been doing about financial mathematics, where you've been given the percentage, to find the profit, here you'll need to calculate the percentage. Remember the key to doing that correctly is to tap into percents' deep connection with fractions and correctly identify both the whole and the part.

One way to talk about profit in terms of percents is the *markup*, which is the gross profit above cost as a percentage of the cost. When calculating the markup, the whole is your cost, and the part is your gross profit. Another way is the profit margin, which is the gross profit above cost as a percentage of the selling price. When calculating the *profit margin*, the whole is your selling price, and the part is your gross profit.

In order to calculate gross profit, markup, and profit margin, there are a couple of formulas you will need to be familiar with:

- Gross profit = selling price − cost

- Markup = gross profit ÷ cost, written as a percentage

- Profit margin = gross profit ÷ selling price, written as a percentage.

**Example:** Suppose you are selling T-shirts. Each shirt costs you $10.14, and you are selling them for $12 each. Calculate the gross profit, the profit margin, and the markup for the T-shirts.

## def•i•ni•tion

**Markup** is gross profit above cost, expressed as a percentage of cost. **Profit margin** is gross profit expressed as a percentage of the selling price.

### Practice Makes Perfect

Problem 12: A computer retailer earns a 15% commission on each computer he sells. If he sells a computer for $1,000, calculate the gross profit, the profit margin, and the markup.

**Solution:** Your gross profit per T-shirt is $12 – $10.14 = $1.86 per T-shirt. The markup is going to be $1.86 ÷ $10.14 = 0.1834. Written as a percent, the markup is 18.34%. The profit margin will be $1.86 ÷ $12.00 = 0.155; written as a percent, the profit margin is 15.5%.

# Percent Error

Every time something is measured, there's a certain amount of error. This becomes obvious if you take a close look at my house. The windows in the dining room are trapezoidal instead of rectangular because someone measured wrong. In the kitchen, the floor and the cabinets don't go all the way to the wall, so you can see light from the basement coming up from behind the kitchen counter. It gives a warm glow at night, but it's definitely not the mark of fine craftsmanship.

**Timely Tips** _____

We use absolute value (see Chapter 1) when calculating the difference between the actual measurement and the accepted value because we don't care which one is bigger and we don't want our percent error to be negative. For example, it doesn't matter whether you over-measure or under-measure a board by 2% of the actual measurement; it's still off by 2%.

The measurement errors in my house are on the order of a fraction of an inch to maybe an inch or two (scientists would prefer if we used centimeters to measure distances). Being off by about an inch is annoying in construction, devastatingly inaccurate in watch-making, and no big deal when measuring the distance to the moon. To give a sense of perspective to the size of the error, we calculate the percent error. In this case, the whole is the accepted or correct value and the part is the difference between the actual measurement and the accepted measurement. To calculate percent error, we find $\frac{|\text{measurement} - \text{actual value}|}{\text{actual value}}$. This percent error calculation only works for numbers. If you measure how cool someone is on the Rad-Awesome-Tubular-Iffy-Ehhh-Ugh-RUN AWAY! scale, you need to invent your own error scale.

**Example:** You're at the carnival and a guy on the midway offers to guess your weight. You weigh 170 lbs and he guesses 160 lbs. What is the percent error of his guess?

**Solution:** We calculate $|160 - 170| ÷ 170 = 10 ÷ 170 = 0.0588$. Expressed as a percent, this is 5.88%.

**Practice Makes Perfect**

Problem 13: The density of water is 1.0 g/cm$^3$. In a science experiment, a student determined it to be 1.2 g/cm$^3$. What is the percent error in the student's measurement?

Problem 14: You get a 500 g bag of peanuts from the supermarket. You measure its weight using a kitchen scale and it weighs 489 g. What is the percent error?

## The Least You Need to Know

◆ Percents are fractions whose denominators are 100.

◆ To convert from a decimal to a percent, move the decimal point two places to the right. To convert from a fraction to a percentage, first convert the fraction to a decimal.

◆ To convert from a percent to a decimal, move the decimal point two places to the left.

◆ The words "percent of" usually mean that you should multiply by the percent written in decimal form.

◆ The words "more than" or "less than" usually mean add (or subtract) your percentage to (from) 100% before multiplying.

◆ Percents are used in many fields, including finance, business, and science.

**6**

# Ratio and Proportion

## In This Chapter

◆ Basing relationships on multiplication and division

◆ Using ratios and proportions

◆ Converting units of currency and measurement

◆ Applying ratios to cooking, science, and other fields

Human beings can have many different relationships: mother, father, spouse, child, aunt, grandfather, friend. As with human relationships, there are many ways two quantities can be related. However, there are two ways that will come up over and over again.

One is like the relationship between my age and my brother's age. I am three years older than my brother. For as long as I've had a brother, he has been three years younger than me, and he always will be three years younger than me. No matter how old we are, if you subtract his age from mine, you will always get three. This is an additive relationship; to get from his age to mine, you add.

Another important connection between quantities is when two numbers are *in proportion*. In this case, the relationship between the two quantities is multiplicative rather than additive. On a baseball diamond, the distance from home plate to second base is always 1.4 times the distance from home plate to first base. This will hold whether you are looking at a little league field or a major league ballpark. If you subtract the distance between home plate and first base from that between home plate and second base, you'll get different numbers on different fields; however, the multiplicative relationship will always hold. We use the word "ratio" to describe this type of multiplicative relationship. Mathematicians would describe this situation as, "the ratio of the two lengths is one-point-four to one." And they would write it, "The ratio of the two lengths is 1.4:1."

# Keeping Everything in Proportion

Let's imagine a math class with 24 students. Suppose that 21 of them say they love math. You might be wondering: Is this an honors math class? Is it the math teacher who was asking people if they love math? Are they promised extra credit? Don't worry about it: 21 of them say that they love math, and they're sticking with their stories. How would you describe the relationship between the students who love math and those who don't?

You might be tempted to say that their relationship is probably difficult or strained or awkward. Especially if there was something a little bit shady about the data collection. Let's rephrase the question: How would you describe the relationship between the *number* of students who say they love math and the *number* of students who don't?

You can talk about how many more people love math than don't. Or you can talk about how many times more students love math than don't. We'll stick with the latter way of looking at things in this chapter. Since there are 24 students in the class and 21 of them love math, then 3 of them don't. Since $21 \div 3 = 7$, there are seven times as many students who love math as who don't. We also can say that the *ratio* of students who love math to those who don't is 21:3. A ratio is when we compare two quantities that have a multiplicative relationship. We use the colon between them to tell us that it is a ratio; the colon functions much like a fraction bar. When we have two ratios that are equal, we call it a *proportion*.

## def•i•ni•tion

A **ratio** between two numbers is a way to express the multiplicative relationship between them. The ratio of water to milk in a recipe is 2:1. This means that for every 1 cup of milk you must have two cups of water, or for every teaspoon of milk you must have two teaspoons of water.

**Example:** There are 60 people at a birthday party and 56 of them eat cake. Write a ratio that describes the ratio between the number of people who eat cake to the number of those who don't.

**Solution:** Because 56 eat cake, 60 - 56 = 4 do not. The ratio of the number of people who eat cake to those who do not is 56:4.

Aside from comparing two quantities in a multiplicative way, ratios can be used to describe situations where one thing scales up to give another. This is commonly seen in architectural drawings, scale models of airplanes, and printing images in graphic design. There will be an original object and a representation of it, often smaller than the original. The relationship between the size of the original and the size of the model will be given in terms of a ratio. Knowing what the ratio is and also knowing the size of the model, you can calculate the size of the original object.

One important concept when dealing with ratios is the idea of a *unit*. A unit is a way of giving physical meaning to otherwise ambiguous numbers. For example, if I say that my nephew weighs 50, you don't know if he is a small child weighing 50 pounds, a growing teenager at 50 kilograms, a baby in the womb weighing just 50 ounces, or if I'm talking about my new semi-truck that weighs a whopping 50 tons! Some typical units you already know about are feet, yards, miles, ounces, pounds, seconds, minutes, hours, days, and years. Even concepts such as students, books, cars, ideas, and moose can be units. You'll learn more about units in science classes if you haven't already.

**Example:** On a blueprint, 2 cm on the drawing represent 300 cm on the ground. What is the ratio of the actual size of the building to the drawing?

**Solution:** The ratio is 300:2.

Let's take that example further. One important quality of ratios is that we can use them in calculations. Ratios allow us to convert measurements from one size to another through multiplication. Replace the colon in the ratio with a fraction bar and perform the division. In this case, we have the ratio 300:2 and we get a *conversion factor* of $\frac{300}{2}$ = 150. If you multiply a length on the blueprint by 150, you will get the length on the ground.

---

### Practice Makes Perfect

Problem 1: Marvin weighs 160 lbs and his elephant friend Green weighs 8000 lbs. What is the ratio of Marvin's weight to Green's weight?

Problem 2: There are 560 ants and 80 people living on the same floor of a building. What is the ratio of the number of ants to the number of people?

**Example:** Using the same information as in the previous example, a wall on the blue-print is drawn as being 3.5 cm long. How long is that wall in the full-size building?

**Solution:** We take the measurement given to us, 3.5 cm, and multiply it by 150 because that is the conversion factor. 3.5 cm × 150 = 525 cm, so the wall is 525 cm long. (You might want to represent this as 5.25 meters.)

# Cooking the Night Away

With the possible exceptions of scientists, students, and others who do quantitative work regularly, there are at least three areas of life where people are most likely to use the math that they learn in school. One of them is when dealing with money, whether it's simple problems based on making change or more complicated calculations about investments. If you travel internationally, you will probably need to exchange currency. The second is dealing with home renovation and construction. Remodeling a house requires careful measurement, the conversion of units, and many other calculations. The third is cooking, especially scaling a recipe to make more or less than it originally calls for.

If you ever cook, you use math. You definitely use fractions as well as proportions. That is, unless you have decided to always follow every recipe exactly as written and never scale it up or down to make a larger or smaller batch.

## Mixed Drinks

You know that powdered lemonade mix, the type where you just add water? Suppose you made two pitchers of lemonade, one where you used three scoops of mix and one where you used four scoops of mix. Which one would taste more lemony? The answer would depend on how much water you added. If you have three scoops of mix and just a little bit of water, it would be much stronger than if you used four scoops of mix and an absolutely huge amount of water. But if both pitchers have the same amount of water, then the one with more powder is going to taste more lemony.

You can capture the lemoniness of the drink by using a ratio. If you can calculate the ratio of mix to water, then you'll be able to compare the relative strength of two pitchers of lemonade.

**Example:** One pitcher of lemonade was made with 3 scoops of mix and 4 pints of water. Another pitcher of lemonade was made with 4 scoops of mix and 5 pints of water. Which pitcher tastes more lemony?

**Solution:** The ratio of mix to water is 3:4 for the first pitcher and 4:5 for the second pitcher. To compare these ratios, convert them to fractions. The one that represents the larger fraction will have more of a lemony taste. 3:4 becomes ¾, and 4:5 becomes ⅘. You need to compare the sizes of these numbers. The easiest way to do that is to convert them to decimals, where they become 0.75 and 0.80. Since ⅘ is a larger fraction, the second pitcher is more lemony than the first.

If you don't like lemonade, you can imagine these problems being done with cocoa mix or any beverage of your choice. This type of reasoning doesn't have to be restricted to problems about tasty beverages. It can be extended to situations involving the vacancy rates of apartment buildings and many other settings.

> **Practice Makes Perfect**
>
> Problem 3: Ford adds 3 spoons of salt into a cup of cola, which has a volume of 500 ml. Arthur adds 10 spoons of salt into a bottle of cola, which has a volume of 2 L. Whose cola is more salty?

## Me Want Cookie!

It's time to raise money for your favorite group, and they're having a bake sale. You've been signed up to make chocolate chip cookies according to their secret recipe; it's always a hit, and the cookies always sell out. They want you to make 300 cookies! Unfortunately, the recipe makes only 48 cookies. What should you do?

You have a few options. You could make 48 cookies, start over and make another 48. Do it all over again and make 48 more, and keep going like this. Eventually you'll have 288 cookies. You can pretend that you ate the other 12. Or you could make another 48 cookies and keep 36 of them for yourself.

This might be a good idea if you're planning on doing your baking over a period of several days—breaking up the work to keep it manageable. Splitting things up also will give you the chance to go back to the store in case you run out of a vital ingredient and need to buy more.

But what if you want to do all of your shopping ahead of time? What if you want to mix up all of your cookies in one giant batch using an industrial mixer? How can you take this recipe for 48 cookies and scale it up to a massive batch of 300 cookies?

While the process of mixing, baking, and cleaning up after 300 cookies might be quite an effort, it's not very hard to rewrite the recipe to make 300 cookies. Once again, this is a problem that requires the use of a proportion. Because you're scaling up the recipe, you're looking for a number we can multiply all the quantities by. You want to multiply 48 by the scale factor to get 300. You'll multiply all of the ingredients by the scale factor.

To find the scale factor for the recipe, you'll take the desired batch size and divide it by the yield of the original recipe. In this case, your scale factor will be 300 ÷ 48 = 6.25. Before you start shopping, measuring, and mixing, you can check your work by multiplying the size of the original recipe by the scale factor and to make sure that you get your desired yield: 48 × 6.25 = 300; since you got your original target number, you know that you calculated correctly.

Now you'll just need to multiply the quantity of every ingredient in the original recipe by 6.25. If the original calls for 1 cup of butter, in your scaled-up version, you'll use 1 × 6.25 = 6.25 cups of butter. If the original calls for 2 cups of sugar, you'll use 2 × 6.25 = 12.5 cups of sugar. If the original calls for $\frac{1}{2}$ teaspoon of salt, your super-size version will use $\frac{1}{2}$ × 6.25 = 3.125 teaspoons of salt.

**Timely Tips**

To find the scale factor for any problem, you'll take the desired size and divide it by the original size. The scale factor can be any positive real number.

**Practice Makes Perfect**

Problem 4: You're making chicken noodle soup. The recipe calls for 4 cups of chopped cooked chicken and 12 cups of water. You decide you want to add 18 more cups of water. How many cups of chicken do you need to make the soup taste the same?

This method also will work in scaling down recipes. If you have a recipe from an industrial bakery and want to scale it down for home use, you will calculate the scale factor the same way. You'll take the quantity you want to make and divide it by the quantity from the original recipe.

**Example:** A domestic diva publishes a recipe for cassoulet in her cookbook, and it makes 100 servings. You want to make this stew for a small soirée and you only need enough to feed 10 people. If the original recipe calls for 14 quarts of chicken stock and 16 pounds of white beans, how much of these ingredients should you use in your scaled-down version?

**Solution:** You want to make 10 servings and the recipe makes 100, so the scale factor will be $\frac{10}{100}$ = $\frac{1}{10}$. To scale down the recipe, you'll multiply every quantity by $\frac{1}{10}$. Therefore, you'll need 1.4 quarts of chicken stock and 1.6 pounds of white beans.

# Unit Multipliers

Some of the most obvious things come in handy again and again. When you learned about the multiplicative identity—the fancy name for the fact that multiplying by one leaves a number unchanged—you probably thought that it was kind of lame. Who cares that multiplying by 1 doesn't do anything?

The trick is to write 1 in a clever way. The sneaky part is that if you take something and divide it by itself you get 1. For example, since 1 foot is exactly the same as 12 inches, $\frac{1 \text{ foot}}{12 \text{ inches}}$ equals 1, because you are dividing something by itself. It doesn't matter that it's written two different ways just as long as it's still the same quantity. This trick is amazing.

### Kositsky's Cautions

Units are a way of specifying a standard of measurement for each quantity. In order to correctly compare ratios, you need to make sure the units are the same in each fraction. If our model bat has a length of 30 inches and the professional bats have a length of 2.5 feet, the ratio between the lengths cannot be 12:1 because 2.5 feet is the same as 30 in.!

You will use this all the time when you are converting from one type of unit to another. In this case, the ratio you set up will be the two different ways to write the same quantity. Any ratio of this type is called a *unit multiplier*. You'll need to do this on a regular basis in science class. At some point, you'll have to convert something from feet and inches to centimeters (or the other direction). If you're ever traveling in another country with a different currency, you'll want to know how much things cost.

### def•i•ni•tion

A **unit multiplier** is any ratio or fraction whose denominator and numerator are equal but in different units. For example, $\frac{1 \text{kg}}{2.2 \text{lbs}}$ and $\frac{€0.73}{\$1}$ are both unit multipliers.

## Show Me the Money: Converting Currency

Let's say that you're from the United States and you are going on vacation to Italy. You're shopping for amazing olive oils and cheeses, fashionable clothes, and European pop music. You're looking to buy a piece of cheese that costs €5. You're wondering how much it costs in United States dollars. Suppose on that particular day, the exchange rate from dollars to Euros is US$1 is equal to 0.73 Euros.

The wrong question to ask here is, "Should I multiply or should I divide?" In all of these problems, you will eventually end up multiplying or dividing, but that's not how you start. The way you start is by setting up a unit multiplier and asking yourself, "Which unit should be in the numerator?" A unit multiplier is what you get when you take the same quantity in two different units and express it as a ratio. In this problem, since $1 is exactly the same as €0.73, there are two ways to make a unit multiplier. You can have $\dfrac{€0.73}{\$1}$ or $\dfrac{\$1}{€0.73}$. There are two choices. Which one should you use in the problem?

In a problem like this, one with only one type of unit, you want the units you are converting *to* in the numerator and the units you are converting *from* in the denominator. Here, you're trying to convert the price of cheese from Euros to dollars, so you should have Euros in the denominator and dollars in the numerator. The unit multiplier that you'll be using is $\dfrac{\$1}{€0.73}$.

**Kositsky's Cautions**

You always need to include the units in a unit multiplier. If you leave the units off, you're not being lazy or efficient or skipping steps, you're being entirely wrong. The only way these work is by having the numerator and the denominator be exactly the same, and for that to hold, you must include the units.

Once you have the correct unit multiplier, the next step is easy. As you might suspect from its name, you will multiply by it. To convert your €5 piece of cheese into dollars, you multiply $€5 \times \dfrac{\$1}{€0.73}$ = $6.85. Just like with multiplying regular fractions (where you can cancel common factors), when you're multiplying this type of fraction, you can "cancel" terms. When the same unit appears in the numerator of one part of the multiplication problem and in the denominator of another part, you can cancel them. You can tell that you've set up one of these problems correctly because the units will always work out. The units that you are converting from will all cancel, and at the end of the problem, you will only have the units that you are converting to.

**Example:** A Canadian is traveling in England and wants to buy a pair of shoes for £75. Suppose that the conversion rate between British pounds and Canadian dollars is £1 = $2.11. How much would the shoes cost in Canadian dollars?

**Solution:** The first step is to set up the unit multiplier. The numerator should be the units that you are converting to, Canadian dollars, and the denominator should be the units you are converting from, British pounds. So you set up the unit multiplier as $\dfrac{C\$2.11}{£1}$.

Next you multiply the quantity that you're converting by the unit multiplier:

$£75 \times \dfrac{C\$2.11}{£1}$ = C$158.25. I hope that those are really nice shoes!

> ### Practice Makes Perfect
>
> Problem 5: Elephant Green is visiting Hong Kong Zoo on a business trip. He wants to buy 5 pounds of bananas in a nearby supermarket for HK$3.30 per pound. Suppose that the conversion rate between Hong Kong dollars and United States dollars is HK$7.79 = US$1. How much would the bananas cost in United States dollars?

# Furlongs per Fortnight: Converting Units

The process of using unit multipliers to convert currency also can be used to convert basic units, like meters to feet. It comes in handy when working with more complicated situations as well. Maybe you're working in international business and trying to find the price in dollars per square foot for putting carpet in your new office space, and you're getting quotes in Euros per square meter. Unit multipliers can be used to solve all of these problems.

**Example:** What is 10 inches in centimeters? The conversion between inches and centimeters is 1 inch = 2.54 cm.

**Solution:** Just like the currency example, you'll need a unit multiplier. Here you're converting from inches to centimeters, so you'll want inches in the denominator and centimeters in the numerator, and your unit multiplier will be $\frac{2.54\,cm}{1\,in}$.

Multiplying, you get $10\ in \times \frac{2.54\,cm}{1\,in} = 25.4$ cm.

**Example:** How much is 10 square inches in square centimeters?

**Solution:** This problem is slightly more complicated because the units are squared. A square inch represents a square that is one inch on a side; a square centimeter is the size of a square that is one centimeter on a side.

Mathematically, we should write "square inch" as $in^2$ and "square centimeter" as $cm^2$. When you're trying to cancel out square inches, you'll need to multiply by the conversion factor *twice* since there are *two* factors of inches that need to be canceled.

Since inches are in the numerator, your conversion factor is again $\frac{2.54\,cm}{1\,in}$.

$$10\ in^2 \times \frac{2.54\,cm}{1\,in} \times \frac{2.54\,cm}{1\,in} = 64.5\ cm^2.$$

This method will work just as well in a problem that has two types of units, such as a quantity represented as a rate. For example, to make sense of a physics problem, you may want to convert a speed from meters per second to miles per hour. You'll need to be a little bit careful in keeping track of which units are in the numerator and which units are in the denominator.

For this type of problem, you'll need to set up your unit multipliers so that the units you don't want cancel out and the units you do want end up in the right places. It's the same idea as before, it's just that there isn't a foolproof rule to make it work.

**Example:** A projectile is traveling at 20 meters per second. How fast is it moving in miles per hour?

**Solution:** In order to solve this problem, you'll need the conversion factors that will allow you to change the distance from meters to miles and that will allow you to change the time from seconds to hours. Typically these are given in a problem; if you're working with a real-world situation, these are very easy to find on the Internet.

> **Timely Tips** _____
>
> The Internet can be a great help in checking your work for this type of problem. If you go to www.google.com and type in "20 meters per second in miles per hour," it will give you the answer 44.7387258 miles per hour.

Here are the conversion factors: 1,000 m = 1 km, 1 mi = 1.61 km, 60 s = 1 min, and 60 min = 1 hr.

The first step in working with a problem where the units are given as a rate, in this case meters per second, is to write the units as a fraction instead of in words. The word "per" should be replaced with a fraction bar. So 20 meters per second will be written as 20 m/s.

Your next goal is to get the meters to cancel out and eventually be replaced by miles. You don't have a direct conversion between meters and miles, so you'll have to go through kilometers. The unit conversion that you'll need will be $\frac{1\text{km}}{1000\text{m}}$. When you set up $20 \text{ m/s} \times \frac{1\text{km}}{1000\text{m}}$, meters will cancel, leaving the distance in kilometers. Next, you'll want to use the conversion between kilometers and miles. To make the km cancel, you'll set up the unit multiplier as $\frac{1\text{mi}}{1.61\text{km}}$.

At this point, your problem will now look like: $20 \text{ m/s} \times \frac{1\text{km}}{1000\text{m}} \times \frac{1\text{mi}}{1.61\text{km}}$.

Don't multiply it out yet. You can cancel the units, eliminating the meters and the kilometers, but don't do anything more that that; don't get involved with the numbers. Now you'll have to take care of the time units. In your problem, as it's set up now, you have seconds in the denominator. To make the seconds cancel, you'll have to put seconds in the numerator of your unit multiplier: $\frac{60\text{s}}{1\text{min}}$. Next, you'll want to cancel the minutes and have your problem in terms of hours, so your last unit multiplier is $\frac{60\text{min}}{1\text{hr}}$.

Putting this all together, you'll have

$$20 \text{ m/s} \times \frac{1\text{km}}{1000\text{m}} \times \frac{1\text{mi}}{1.61\text{km}} \times \frac{60\text{s}}{1\text{min}} \times \frac{60\text{min}}{1\text{hr}}.$$

Now you can multiply everything out, canceling all the units, leaving you with 44.74 mi/hr.

---

### Practice Makes Perfect

Problem 6: In Argentina, there are 2 cows for each person. In New Zealand, there are 12 sheep for each person. Assume sheep weight about 70 kg each, and cows weight about 400 kg each. If these are all the livestock in those countries, which country has more weight of livestock per person?

---

## The Least You Need to Know

- ◆ Ratios are written with a colon separating the two numbers being compared in a multiplicative relationship.

- ◆ To calculate with ratios, you will often replace the colon with a division sign or fraction bar.

- ◆ To convert units, multiply by unit multipliers in a way so that the units you don't want cancel out and the units you do want are introduced.

- ◆ Ratios can be used to solve a variety of problems around the house and in science.

# Part 3

# Radical and Exponential Power

Maybe the image of a "radical" conjures up someone to watch out for, someone who is out to cause trouble? That fits, somewhat, as mathematical radicals tend to trip some students up.

Don't worry, these ideas aren't only out to cause trouble. This part also shows you the mathematics of population growth, the growth of bacteria, and the spread of rumors and diseases. It's also the source of the calculations that let you get serious about money. Not only is this one of the sources of mathematical models in the social sciences, financial realm, and public health, but it's also your chance to work with impressively large numbers.

# Chapter 7

# Integral Exponents

## In This Chapter

- ◆ Positive integral exponents
- ◆ Order of operations with exponents
- ◆ Expressing integers in exponential notation
- ◆ Negative integral exponents

There's a story in the mathematical folklore about a king and one of his ministers (the "government official" type of minister, not the "spiritual leader" type). The minister had given the king some advice, maybe helped him crush an uprising, and the king wanted to thank him.

The king asked the minister what would be a fair reward. The minister replied that he'd take some gold coins, and proposed the following method of deciding how many coins he'd take. They would take a chessboard and put 1 coin on the first square, 2 coins on the second square, 4 coins on the third square. The number of gold coins would keep doubling on each square, continuing for all 64 squares of the chessboard. The minister would receive the coins in the last square.

The king, being quite skilled at multiplication and at working with exponents, thought for a moment after hearing this plan, and decided to have the greedy minister killed. If you can't see why he made this decision, after you read this chapter, try to calculate $2^{63}$.

You can get a sense of the power of successive doubling if you try to fold a piece of paper in half eight times. If you try with regular paper—or with just about any piece of paper you find lying around without having to make a special order—it would be difficult or impossible to do so.

The record for folding any piece of paper in half is 12 times, done in December of 2001 by a high school student named Britney Gallivan who used a special piece of industrial toilet paper. The final folded piece of paper was over 2,000 times thicker than the original! If you took a regular sheet of paper and wanted to fold it in half 50 times, you couldn't do it for any number of reasons, including the fact that your piece of folded paper would be over 200,000,000 kilometers (130,000,000 miles) tall or longer than the distance to the sun.

With exponents we often end up with large numbers and with numbers incredibly close to zero. They also provide us with the tools for working with very large and very small numbers.

# The Power of Exponents

This is the last piece of the arithmetic puzzle that fits together to be pre-algebra. Once you learn about exponents, you'll know all of the ways that numbers are combined in all of pre-algebra, and you'll have the basic tools for solving all of its problems. You still have to learn the rules for working with exponents and methods of solutions of the problems, but you now have a fighting chance.

Exponents are merely a shorthand method to work with lots and lots of multiplication. Because you've already learned a lot about multiplication, you shouldn't have much trouble generalizing it to the case of exponents.

## Repeated Multiplication

The easiest use of exponents is as a shortcut for multiplying the same number by itself a bunch of times. Instead of writing $5 \times 5 \times 5 \times 5 \times 5 \times 5 \times 5 \times 5 \times 5 \times 5 \times 5$, we can use the much smaller and more efficient notation of $5^{11}$. The normal-looking 5 is called the base, and it's the number that you're going to multiply by itself. The tiny 11 in the upper right tells you that you're going to multiply 5 by itself 11 times. The expression $5^{11}$ is read as "five to the eleventh" or "five to the eleventh power."

**Example:** Write $1.8 \times 1.8 \times 1.8 \times 1.8 \times 1.8 \times 1.8$ in exponential form.

**Solution:** $1.8^6$.

**Example:** Use repeated multiplication to expand $11^4$.

**Solution:** $11 \times 11 \times 11 \times 11$.

Back in Chapter 5, you worked with financial problems involving compound interest and were told these problems are easier to do when taking advantage of several rules of math including exponents. To calculate the amount of money in your savings account after a certain number of years, take the interest rate written as a decimal, add 1 to it, and then raise it to the exponent of the number of years. Finally, multiply this number by the initial balance in the account.

**Timely Tips** _____

Any number to the first power is itself. So $6^1 = 6$ and $234,987^1 = 234,987$. Because both forms are equivalent, it's common to leave off the exponent of 1; however, it's sometimes included when someone is trying to make a point and emphasize that the number is being written in exponential form.

**Example:** If you put $1,200 in an account earning 4% interest and leave it in the account for 30 years, how much will be in the account?

**Solution:** 4 percent written as a decimal is 0.04. Adding 1 you get 1.04, which is the base of the exponent. Next calculate 1.04 raised to 30, the number of years: $(1.04)^{30} =$ 3.24339751 …. Finally, multiply this by the initial balance. $1200 \times 3.24339751 … =$ $3892.077012 ….

Because this is a financial problem, we round to the nearest cent at the end of the problem, giving us $3892.08. Your money has more than tripled, and you didn't need to do anything other than sit around and wait!

You also can use exponents to solve problems involving populations. Typically these are the infamous bacteria problems. You'll be told how many bacteria there are in the Petri dish at the beginning of the experiment and how long it takes for their population to double. Or you might be given the population of a city and be told what factor the population is multiplied by each year and be asked to find the population in a given year.

**Example:** A scientist has 3 bacteria in a Petri dish, and the population of bacteria doubles every 6 hours. How many bacteria are in the dish after 48 hours?

**Solution:** In 48 hours, there are eight 6-hour periods, so the population will double eight times. The number of bacteria would then be given by $3 \times 2^8 = 768$.

**Example:** A city starts with 20,000 people. Each year the population of the city grows by 20 percent. How many people live in the city after 10 years?

**Solution:** Growing by 20% in a year implies that at the end of a year, there are 1 + 0.2 = 1.2 times as many people as there were in the beginning of the year. Thus, after 10 years 20,000 × $(1.2)^{10}$ people = 123,835 people will live there.

**Example:** A city starts with 80,000 people. Each year the population of the city decreases by 10%. How many people live in the city after 15 years?

**Solution:** Shrinking by 10 percent in a year implies that at the end of a year, there are 1 − 0.1 = 0.9 times as many people as the beginning of the year. 80,000 × $(0.9)^{15}$ people = 16,471 people.

---

### Practice Makes Perfect

Problem 1: In the story about the king and the minister mentioned at the beginning of this chapter, how many coins would the king have to put on the tenth square, if the king had accepted the minister's method?

Problem 2: A monster eats 500 copies of *The Complete Idiot's Guide to Pre-Algebra* on the day he is created, and the number of copies of the book he eats triples every day. How many copies does he eat when he is 15 days old?

---

## Writing Integers in Exponential Form

Back in Chapter 1 when you factored numbers into products of primes, you wrote them out the long way using repeated multiplication. You probably didn't think of it as repeated multiplication at the time—you probably just thought that it was inefficient. It's usually preferable to use the more compact exponential notation when writing a number in its factored form.

### Practice Makes Perfect

Problem 3: Factor 86,400,000 into a product of primes and write the answer in exponential form.

**Example:** Factor 360 into a product of primes and write the answer in exponential form.

**Solution:** You can make a factor tree to find that
$360 = 2 \times 2 \times 2 \times 3 \times 3 \times 5 = 2^3 \times 3^2 \times 5$.

# Order of Operations Reloaded

Now that you have another way of working with numbers, you need to know how it fits together with the other ways in the order of operations. If you're faced with an expression such as $\dfrac{17+9\left[-5^2+(18\times 9)-3\right]^{17}}{265-9\left[\left(1.148-2.91^2\right)+8^4+(-3)^2\right]^{17}}$, you'll need to know how to evaluate it. Here are the complete rules of the order of operations:

1. Calculate quantities in parentheses first. If there are nested sets of grouping symbols (like parentheses or brackets), start with the innermost set.

2. There are invisible parentheses around each of the following: the numerator of a fraction, the denominator of a fraction, and an exponent.

3. After calculating things in parentheses, next evaluate exponents.

4. Once you've dealt with the exponents, do all multiplications and divisions from left to right.

5. Finally, do additions and subtractions from left to right.

**Example:** Use the order of operations to simplify $\dfrac{\left[(6-3)\times 2\right]^2-(4-1)^3}{6-4}$.

Solution: You need to start by doing the calculations inside the grouping symbols, starting with the innermost sets first. This gives you $\dfrac{\left[(3)\times 2\right]^2-(3)^3}{2}$. Remember that there are invisible parentheses around the denominator, so you calculate the 6 − 4 in the denominator in this first step.

There are still more grouping symbols, so the next step is to calculate 3 × 2, and the problem becomes $\dfrac{\left[6\right]^2-(3)^3}{2}$.

**Timely Tips**

Some people remember the order of operations as PEDMAS: parentheses, exponents, division, multiplication, addition, subtraction.

**Kositsky's Cautions**

One very common mistake is to group a negative sign or minus sign outside parentheses with a number when that number is raised to an exponent. This is wrong. For example, $-4^2$ means $-(4)^2$, which is equal to $-16$, not 16! Unless a minus sign is written inside parentheses, always assume that it is not affected by the exponent.

That takes care of most everything. Next you should do the exponents, giving you $\frac{36-9}{2}$.

Almost done! Subtract in the numerator to get $^{27}/_2$. This is a perfectly fine fraction, but maybe you prefer mixed numbers. In that case, you can write it as $13\frac{1}{2}$ or 13.5.

---

### Practice Makes Perfect

Problem 4: Calculate $\dfrac{-5\times\left[3.14-\left(-3.14\right)^{3}\right]}{\left(3^{2}-2^{3}\right)}$.

Problem 5: Calculate $\dfrac{-2.1\times\left(-3^{5}+\left(-4\right)^{4}\right)-\left(-1\right)^{1}}{2.5-2.25^{2}\times3^{4}}$.

---

# Multiplying with Same Base

The rules for multiplying exponents with the same base are easy once you get used to them, which is how I got away without memorizing them for so long. If you multiply a number by itself a bunch of times and then multiply it by itself some more, it's not that hard to figure out how many times it has been multiplied by itself. It's mostly an issue of counting.

## Positive Bases

Let's say you have $5^3$ and you multiply it by $5^4$. That would be $5 \times 5 \times 5$ multiplied by $5 \times 5 \times 5 \times 5$, which equals $5 \times 5 \times 5 \times 5 \times 5 \times 5 \times 5 = 5^7$. Do you see the pattern here? $5^3 \times 5^4 = 5^{3+4} = 5^7$. When you multiply two exponentials with the same base, you add the exponents. If you forget how the rule works, you can always write out the repeated multiplication and count how many you have of each factor.

**Example:** Write as a single exponential: $7^{89} \times 7^{167}$.

**Solution:** In this type of problem, it would be long and tedious to write out 89 copies of 7 followed by 167 more copies of 7—and you might make a mistake counting them all up, too. Here's when it's easiest to use the rule: you just add the exponents, so the answer is $7^{89 + 167} = 7^{256}$.

You will use this again and again and again. It might come up in different settings, but it's still the same method, even if you have to use a few rules of multiplication to sort things out.

**Example:** Use the rules of multiplication and exponents to simplify $3^6 \times 7^4 \times 3^3 \times 2^5 \times 7$.

**Solution:** The only rule that we have so far is simplifying the product of exponents that have the same base. In this problem there are three different bases: 2, 3, and 7. Fortunately, we can use the commutative and associative laws from Chapter 1 to rewrite this problem as: $2^5 \times (3^3 \times 3^6) \times (7 \times 7^4)$. Now we can apply the rule for multiplying exponentials with like bases to the part with the 3s and to the part with the 7s, and our overall answer is $2^5 \times 3^{3+6} \times 7^{1+4} = 2^5 \times 3^9 \times 7^5$.

This covers almost every case involving multiplying exponentials. There's one more situation that can't be talked about until the end of this chapter. How could you use the rules of exponents to simplify $2^6 \times 4^2$? With the rules the way that we've stated them, it doesn't seem like you have many options. The bases are different: 2 and 4 are not the same number, so the rule shouldn't apply. However, $2^6 \times 4^2 = 1,024$, which can be written as both $2^{10}$ and $4^5$. We'll return to this type of example at the end of the chapter, after you've learned how to raise an expression with an exponent to another power.

## Negative Bases

At some point, you or your teacher probably will want to know if some expression is positive or negative. If you only have positive numbers and no subtraction signs, this is really easy! Especially after bringing in exponents, this kind of problem can get tricky. We'll go through a few examples of negative numbers raised to powers to teach you how to do it like a pro.

**Example:** Is the following expression positive or negative: $(-3)^4$?

**Solution:** Exponentiation is simply repeated multiplication, so we know that $(-3)^4 = (-3) \times (-3) \times (-3) \times (-3)$. The associative property says we can do these multiplications in whatever order we choose, so we'll multiply the first two terms together and the last two terms together, like this $[(-3) \times (-3)] \times [(-3) \times (-3)]$.

$(-3) \times (-3) = 9$ so our problem becomes $[9] \times [9] = 81$. Therefore, $(-3)^4$ is a positive number.

The pairing we did with square brackets wasn't very hard because there were only 4 terms. However, sometimes the exponent is very, very large and you can't—or at least I can't—write down hundreds or thousands of –3s. You'd probably either die of boredom or lose count at some point. Because the exponent is even, $(-3)^4$ is positive. In fact, all real numbers with even exponents are positive! But why would you take our word for it? Probably because we're writing this book—but that's beside the point. I suppose it would be more accurate to say that "point" is beside the point, but let's keep that between us.

When you raise a negative number to an even power, your answer is always a positive number because you can pair each multiple with another and have none left over. When you raise a negative number to an odd power, your answer is always a negative number because there will always be one negative number left all by its lonesome.

In other words, with an even number of factors you can use the Associative Law to group everything into pairs, like we did with the four –3s from the previous example. This can continue with as many pairings as you want. The reason odd powers of negatives are negative is that there is one negative number without a partner. You can see him in the next example.

**Example:** Is the following expression positive or negative: $(-7)^{6,357}$?

**Solution:** Yikes! It's going to take way, way too long to write out 6,357 copies of –7. And if you miscount, you may get the wrong answer. I certainly can't keep track of 6,357 identical figures written down!

Let's try to think about the pairings together. To make it simple, we're going to pair neighboring numbers with the odd number first and count them as we go along.

$$\underbrace{\left[(-7)\times(-7)\right]}_{1 \text{ and } 2}\times\underbrace{\left[(-7)\times(-7)\right]}_{3 \text{ and } 4}\times\underbrace{\left[(-7)\times(-7)\right]}_{5 \text{ and } 6}\times\underbrace{\left[(-7)\times(-7)\right]}_{7 \text{ and } 8}\times\ldots\times\underbrace{\left[(-7)\times(-7)\right]}_{6355 \text{ and } 6356}\times\underbrace{\left[(-7)\right]}_{6357}$$

We skipped writing all the groupings in the middle, but they're still there. Remember, the ellipsis means you continue the same pattern. Since $(-7) \times (-7) = 49$, let's replace all of the pairs of –7s with 49.

$$\underbrace{[49]}_{1 \text{ and } 2}\times\underbrace{[49]}_{3 \text{ and } 4}\times\underbrace{[49]}_{5 \text{ and } 6}\times\underbrace{[49]}_{7 \text{ and } 8}\times\ldots\times\underbrace{[49]}_{6355 \text{ and } 6356}\times\underbrace{\left[(-7)\right]}_{6357}$$

Since 49 is positive, you can multiply 49 by itself to your heart's content and you will never get a negative number. The vengeful (–7) who doesn't have a partner ruins the party for everyone and switches the sign. A negative number times a positive number (even a really big one like $49^{3,178}$) is negative. This means $(-7)^{6,357}$ is negative.

While writing all those curly braces was fun, manually pairing up numbers is time-consuming when you're doing more than about 6. Instead of writing all that out, we can use the trick about negatives raised to odd powers and give a quick, simple, and correct answer. Because the exponent is *odd*, by the rule given above $(-7)^{6,357}$ is *negative*. See how easy that is? Now all you have to do is memorize the rule. If you don't understand why the rule works, don't worry. Understanding will come in time.

## Can You Make an Exception?

It seems like there's always a catch in math. There are situations when general principles don't always apply. It's as if there's someone waiting to jump out and yell "gotcha!" To be perfectly honest, there is one very special case where adding two expressions with exponents is secretly like multiplying. Crazy, I know. If this is too weird for you, just skip the rest of this section; you can get away without really knowing it. It doesn't come up very often, and there are other ways around it.

Here's the special case: Remember how multiplication was just repeated addition? So if you have $17 + 17$, it's really just $2 \times 17$. (If you add a number to itself, that's the same as two times the number, right?) Same thing applies to more complicated numbers, so $2^7 + 2^7$ would be the same thing as $2 \times 2^7$. And is now the product of two exponentials with the same base, 2. So $2^7 + 2^7 = 2 \times 2^7 = 2^1 \times 2^7 = 2^8$.

For the same thing to work with 3s, you'll need 3 copies.

> ### Practice Makes Perfect
>
> Problem 8: Is $(-9)^{2,357}$ positive or negative?
>
> Problem 9: Interpret addition as repeated multiplication to simplify $5^4 + 5^4 + 5^4 + 5^4 + 5^4$.
>
> Problem 10: Interpret addition as repeated multiplication to simplify $(-6)^9 + (-6)^9 + (-6)^9 + (-6)^9 + (-6)^9 + (-6)^9$.

**Example:** Interpret addition as repeated multiplication to simplify $3^{10} + 3^{10} + 3^{10}$.

**Solution:** $3^{10} + 3^{10} + 3^{10} = 3 \times 3^{10} = 3^{11}$.

# Negative Exponents

There's no rule saying that exponents need to be positive. Mathematically, it's just fine to have an expression like $2^{-6}$. You might be wondering how you could multiply a number by itself a negative number of times. You might suspect that this is some sort of mathematical trick, a way to trip you up with numerical crazy-talk that makes no sense at all.

Don't despair! Negative exponents might seem weird at first, but they have a fairly simple explanation. Negative exponents are a short-cut for repeated division. Division, you might ask? It makes sense: since division is the same as un-multiplying (remember the inverse operations from Chapter 3?), multiplying a negative number of times would be un-multiplying—in other words, dividing.

We rarely express negative exponents as division, though; instead, it's usual to write them in terms of fractions or reciprocals. Remember that dividing by a number is the same as multiplying by its reciprocal, so repeated division can be rewritten as repeated multiplication by its reciprocal.

**Example:** Expand $3^{-6}$ as a product of reciprocals.

**Solution:** The reciprocal of 3 is $\frac{1}{3}$, so $3^{-6}$ is $\frac{1}{3} \times \frac{1}{3} \times \frac{1}{3} \times \frac{1}{3} \times \frac{1}{3} \times \frac{1}{3}$.

A negative exponent also can be interpreted as the reciprocal of the same base but with a positive exponent.

**Example:** Express $3^{-6}$ as the reciprocal of 3 raised to a positive exponent.

**Solution:** $(\frac{1}{3})^6$.

You can also use negative exponents to simplify a product when you have repeated multiplication in the denominator.

**Example:** Use a negative exponent to simplify $\dfrac{1}{6 \times 6 \times 6 \times 6}$.

**Solution:** Since there are four copies of 6 in the denominator, this can be simplified as $6^{-4}$.

**Example:** Use a negative exponent to simplify $\dfrac{2}{7 \times 7 \times 7}$. In this case, the numerator is not 1, so you first need to factor out the numerator to write this in the form that we've been working with: $\dfrac{2}{7 \times 7 \times 7} = 2 \times \dfrac{1}{7 \times 7 \times 7} = 2 \times 7^{-3}$.

---

### Practice Makes Perfect

Problem 11: Simplify $\frac{1}{8} \times 8^5 \times 8^3$.

Problem 12: Simplify

$$\dfrac{1}{(-7)^{13}} \times (-7)^5 \times (-7)^5 \times (-7)^5 \times (-7)^5 \times (-7)^5 \times (-7)^5 \times (-7)^5 \times (-7)^{21}.$$

Now you can work with both positive exponents and negative exponents. Is there anything left? Or have you mastered every type of exponent? Exponents still have a few tricks up their sleeve. In Chapter 9, you'll learn about fractional exponents. (Really, we're not making that up.) There is one more type of exponent to deal with now: zero as an exponent.

With zero as an exponent, you're not multiplying any times, and you're not dividing any times. What should the answer be? We declare that the answer is always 1. No matter what number you think of, when you raise it to the zeroth power, the answer will be 1.

**Example:** What is $6^0$? $294,587^0$? $(-239,847)^0$?

**Solution:** They are all 1.

**Timely Tips** _____

Any number to the zeroth power will give an answer of 1. Even $0^0 = 1$. Explaining why $0^0 = 1$ requires calculus, which at this point is the mathematician's way of saying "because I said so."

# Dividing with Same Base

Lots of ideas in math come in pairs. It's both good and bad. In some sense it means that you only need to learn half as many things because learning a pair of things is easier than learning two separate things that don't go together at all. Unfortunately, you also run the risk of getting things backward.

Multiplication and division go together much the same way that addition and subtraction go together. And we can exploit that relationship when thinking about division of exponentials with the same base. When you divide exponentials with the same base, you subtract the exponents. Multiplying, you add; dividing, you subtract.

Simply apply the rule to calculate $11^8 \div 11^3 = 11^5$. To see why this works, we'll be best off returning to the realm of fractions. We can imagine the division problem $11^8 \div 11^3$ as the fraction $\dfrac{11^8}{11^3} = \dfrac{11 \times 11 \times 11 \times 11 \times 11 \times 11 \times 11 \times 11}{11 \times 11 \times 11}$. Dividing out, we get

$$\frac{\cancel{11} \times \cancel{11} \times \cancel{11} \times 11 \times 11 \times 11 \times 11 \times 11}{\cancel{11} \times \cancel{11} \times \cancel{11}} = 11^5.$$

A slightly trickier variant on division of exponentials would be to ask $5^4 \div 5^{10}$. Again, it's a division problem where both numbers have the same base, but this time the second number has the higher exponent. Either way

| **Practice Makes Perfect** |
| --- |
| Problem 13: Calculate $13^2 \times 13^3 \times 13^5 \div 13^4 \div 13^2$. |

of thinking about division of exponentials will work here: you can either apply the rule and subtract the exponents or else you can set up a fraction and divide out the common factors.

If you apply the rule, you'll have $5^4 \div 5^{10} = 5^{(4-10)} = 5^{-6}$. Approaching the division problem as a fraction, you'd have

$$\frac{5\times5\times5\times5}{5\times5\times5\times5\times5\times5\times5\times5\times5\times5} = \frac{\cancel{5}\times\cancel{5}\times\cancel{5}\times\cancel{5}}{\cancel{5}\times\cancel{5}\times\cancel{5}\times\cancel{5}\times5\times5\times5\times5\times5\times5}.$$ This gives you

the same answer: $5^{-6}$.

# More Confusing Matters with Bases

Sometimes you'll come across like bases that are hidden. This type of problem is similar to playing baseball with the bases thrown randomly around the field. If first base is glued to third and second is in the bleachers, someone needs to collect them, separate them, figure out which is which, and put them where they're supposed to be.

Whenever a number is raised to a power, you can put parentheses around the number like this: $16^2 = (16)^2$.

This may seem pointless, and sometimes it is. However, if your number is not prime, it allows you to forget about the exponent for a while and do the activity you love most: factoring! $(16)^2 = (2^4)^2$

Raising a number written with an exponent and a base to some power is actually quite simple. Multiply the exponent by the power and you have your new exponent. Keep the base the same. You're finished. $(2^4)^2 = 2^{4\times2} = 2^8$

Luckily, it's easy for you to check your answer in this case. $16^2 = 256$ and $2^8 = 256$, so the expressions are equal.

---

**Practice Makes Perfect**

Problem 14: Simplify $4^{-2} \times 16^3 \times 8^{-6} \times (-1{,}024)^2$.

---

We're done ungluing the bases, but now we need to put them back in their places. To do this we need to factor the base into its prime factorization, multiply exponents as needed, and combine the powers of like bases.

**Example:** Write $25^3 \times 4^{-5} \times 2^2 \times 5^2$ as the product of primes raised to powers.

**Solution:** First put parentheses around each non-prime. If you're not sure if a number is prime, put parentheses around it.

$(25)^3 \times (4)^{-5} \times 2^2 \times 5^2$

Now rewrite the composite numbers as the product of primes.

$(5^2)^3 \times (2^2)^{-5} \times 2^2 \times 5^2$

Because $5^2$ and $2^2$ are raised to a power, you should multiply their exponent by that power and jot down the new expression, same bases with new exponents.

$5^{2 \times 3} \times 2^{2 \times (-5)} \times 2^2 \times 5^2$

$5^6 \times 2^{-10} \times 2^2 \times 5^2$

Finally, we can combine the like bases by adding their exponents like we did earlier in this chapter.

$5^{6+2} \times 2^{-10+2}$

$5^8 \times 2^{-8}$

We can now play with these primes easily! There are only two so I guess we'll have to play catch.

Whoever glued first and third bases together has been at it again. This time, he's glued the bat to the base and hid them beneath home plate. We can use the same basic strategy to solve problems with composite bases (or bats) by distributing the power to the exponent of each of the primes in the prime factorization.

**Example:** Write $60^{20}$ as the product of primes.

**Solution:** First we rewrite 60 in its prime factorization: $2^2 \times 3 \times 5$ giving us $(2^2 \times 3 \times 5)^{20}$.

You already know how to distribute the power if there's only one factor in the base, and having multiple factors doesn't change the method. Multiply each of the exponents by the power to get rid of that 20.

$(2^{2 \times 20} \times 3^{1 \times 20} \times 5^{1 \times 20})$

$2^{40} \times 3^{20} \times 5^{20}$

> **Practice Makes Perfect**
>
> Problem 15:
> $(-8)^4 \times 9^2 \times 16^{-3} \times 27^4$
>
> Problem 16:
> $3^9 \times (-5)^{12} \div (-1/81) \times 0.2^{-4}$

**Kositsky's Cautions**

You cannot distribute the outer exponent if there is anything other than multiplication, division, or exponents within the parentheses. You must evaluate all of the addition and subtraction in the parentheses beforehand.

> ### Practice Makes Perfect
>
> Problem 17: Write 210 as the product of primes.

There are only primes with exponents remaining so we're finished.

Sometimes you'll come across bases that are raised to the same power. These are grouped together as naturally as peas and carrots ... to the fourth power. If you have some expression such as $5^4 \times 7^4$, then you can combine terms as $(5 \times 7)^4$.

**Example:** Write $5^5 \times 6^4 \times 13^5 \times 2 \times 3$ with as few bases as possible.

**Solution:** First we rewrite this product in its prime factorization: $5^5 \times (2 \times 3)^4 \times 13^5 \times 2 \times 3$.

We distribute the exponent over $2 \times 3$ and gather the rogue 2 and 3 from the last two terms in the product. This gives us $5^5 \times 2^5 \times 3^5 \times 13^5$.

> ### Practice Makes Perfect
>
> Problem 18: Write $6^3 \times 2^2 \times 63 \times 7^6$ with as few bases as possible.
>
> Problem 19: Write $(-2)^4 \times 6^2 \times 70 \times 22 \times 3^6 \times 35^5 \times 11^5$ with as few bases as possible.

Notice that all the bases are raised to the same exponent, so we can multiply the bases together and raise them to their common exponent, 5. This gives us our final answer of $390^5$.

Because there is nothing left but one base (you can't have fewer than that), we're done! With this chapter, too. Okay, you do have a few practice problems to complete.

## The Least You Need to Know

- Positive exponents tell you how many times you multiply a number by itself.

- Exponents come before multiplication, division, addition, and subtraction in the order of operations.

- Negative exponents tell you how many times to multiply the reciprocal of a number by itself.

# Roots and Other Exponents

## In This Chapter

- ◆ Square roots
- ◆ Other roots
- ◆ Fractional exponents and simplifying them
- ◆ Multiplying and dividing expressions with exponents
- ◆ Raising an exponent to another power

Ideally you should learn the rules of exponents, internalize them, and be able to apply them automatically. Same goes for all the rules of math. You shouldn't need to think about them at all; you should just do them, as if your pencil was guided by the invisible hand of mathematics, leading you swiftly and accurately to the correct answers to your math problems.

It doesn't always work like that. I'm going to let you in on a secret: I didn't memorize the rules of exponents until I was an adult. Not only had I completed a math major and finished a Ph.D. in math without memorizing the laws of exponents, but I had been teaching math for several years. I can't recommend that approach. You will be able to solve problems much more quickly and efficiently if you learn the rules of exponents.

You can't get something for nothing, and there was a trade-off in my inability to remember the rules of exponents. Every time I did a problem with exponents, I had to think about their properties and write things out the long way in order to figure out what the right answer should be. I was still able to answer the problems, but I couldn't do them quickly and effortlessly. I didn't make up my own rules or guess at what the answer might be—I had to figure it out from basic principles each time I did a problem. It made me a lot slower and more error-prone when I did problems in algebra and calculus classes that involved logarithms.

Only after teaching calculus every semester for several years did the rules eventually stick. It's amazing how much easier basic algebra problems have become since I've learned the rules of exponents.

Learn from my mistakes and learn both the rules of exponents as well as the underlying principles that make them work. You'll be able to focus your attention on the big picture of the problem without having to waste time worrying about the picky details.

As exponents are a short-cut way of dealing with multiplication, and multiplication and division are related, it should come as no surprise that most of the rules of exponents have to do with multiplication, division, and exponentiation. In fact, when you see expressions with exponents added or subtracted, there is very, very, very rarely anything that you can do to further combine them or simplify them. There are no general rules for adding and subtracting exponents. So if you see something like $3^6 + 3^7$, there is probably nothing that you can do to further simplify the problem.

**Amy's Answers**

Abstraction is one of mathematicians' most treasured methods. Bigwig mathematicians are generally not satisfied with any mathematical idea or concept that hasn't been abstracted so completely that it no longer exists in the real world.

In this chapter, you'll learn about square roots and other roots, and their relationship to exponents. Once you've finished this chapter, you should know every operation you will ever need to inflict on a number in this course. From here on out, it's learning how to use these numerical tools in a variety of situations, including abstraction and word problems.

# Defining Square Roots

What positive number do you multiply by itself to get an answer of 9? That's not hard to figure out as 9 is a relatively small number, and $3 \times 3$ should come to you fairly quickly. We can ask the same question about larger numbers, like 100 (you should be thinking $10 \times 10$). There's no reason to stop there; we can ask this question about any number that we want to.

You've probably noticed that mathematicians are not a wordy bunch, and whenever there is an idea defined it quickly is assigned a short name and often a cryptic-looking notation. It's no different in this case. The question "What number do you multiply by itself to get …?" is rephrased in mathematical language as "What is the square root of …?" The answer to the question is known as the *square root*. The notation for square root is a funny-looking symbol that somewhat resembles a division sign and that looks like $\sqrt{\phantom{x}}$ and is called a "radical." You put the number that you're looking for the square root of under the radical. To translate our opening question of "What number do you multiply by itself to get an answer of 9?" into the language of math, we'd write, "What is $\sqrt{9}$ ?"

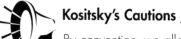

**Kositsky's Cautions** _____

By convention, we allow only positive numbers to be square roots of other positive numbers. For example $(-3)^2 = 9$, but because $-3$ is negative, $-3$ is not the square root of 9.

Every now and then in math you run into something that's forbidden, like not being able to divide by zero. There's one thing that you can't do with square roots. You can't take the square root of a negative number. The reason behind this is that when you take a real number and multiply it by itself, the answer is always positive. You can never get a negative number as an answer when you multiply a real number by itself.

# def•i•ni•tion _____

The **square root** of a positive number is a second positive number such that the second number squared is equal to the first. For example, $\sqrt{25} = 5$.

**Kositsky's Cautions** _____

You should be on your guard if you encounter the square root of a negative number. At the level of pre-algebra, this type of problem can never be solved. Higher math has a loophole that lets these types of problems weasel through. At this point, you should be very, very wary if you see a negative number under a radical sign.

# Calculating Square Roots

Looking at a problem like $\sqrt{9}$ or $\sqrt{16}$, it isn't so hard to figure out the answer. The numbers are fairly small. There aren't a lot of numbers to check. The answers are fairly easy to guess. For a lot of these types of problems, one of the best strategies is to just memorize some of the most common square roots.

**Example:** What are $\sqrt{1}$, $\sqrt{4}$, $\sqrt{9}$, $\sqrt{16}$, $\sqrt{25}$, $\sqrt{36}$, $\sqrt{49}$, $\sqrt{64}$, $\sqrt{81}$, and $\sqrt{100}$?

**Solution:** The answers are 1, 2, 3, 4, 5, 6, 7, 8, 9, and 10. You should memorize these.

Working with other numbers requires a few other techniques ranging from factoring to using a bit of guess-and-check.

## Perfect Squares

So far every problem in this section has had a whole number answer. That hasn't been a coincidence; the numbers have been carefully chosen to come out as whole numbers. If you take the square root of a number and get a whole number as an answer, then your original number is called a perfect square. The numbers 1, 4, 9, 16, 25, … are all perfect squares because their square roots are whole numbers.

*Perfect squares.*

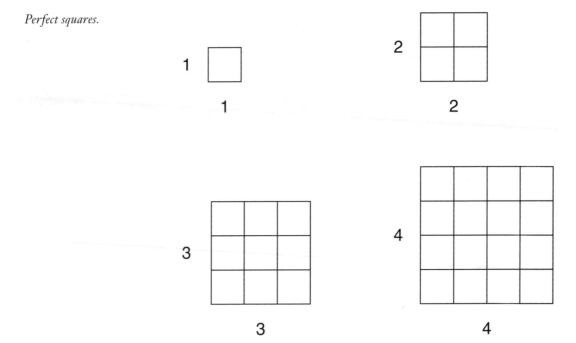

Unfortunately, unless you're faced with a number that you've memorized as a perfect square, you won't know whether or not a number is a perfect square until after you've finished the problem. Therefore, the first step in taking a square root doesn't really matter, and it's the same for every whole number. You factor it into a product of primes.

Once you've written your number as a product of primes in exponential notation, look at the exponents. If the exponents are all even, then your number is a perfect square. If one or more of the exponents is odd, then your number is not a perfect square. Remember that if there is no exponent written, there is still an exponent of 1—an odd number.

**Example:** Determine whether or not 1,200 is a perfect square.

**Solution:** Factoring, $1,200 = 2^4 \times 3 \times 5^2$. The exponents on the factors of 2 and 5 are even, but the exponent on the 3 is 1, which is an odd number, so 1200 is not a perfect square.

**Example:** Determine whether or not 14,400 is a perfect square.

**Solution:** Factoring, $14,400 = 2^6 \times 3^2 \times 5^2$. All of the exponents are even numbers, so 14,400 is a perfect square.

Once you have a number factored and know that it's a perfect square, it's not much more work to find its square root. The square root will have the same prime factors, but you halve all of the exponents.

**Example:** Find $\sqrt{14400}$.

**Amy's Answers**

It's really easy to check your answer for finding square roots. Simply take your answer and multiply it by itself. If it's the same as the radicand, your answer is correct. If not, you've made an error somewhere along the way.

**Practice Makes Perfect**

Problem 1: Evaluate $\sqrt{121}$.

Problem 2: Evaluate $\sqrt{\dfrac{1}{36}}$.

**Solution:** Because $14,400 = 2^6 \times 3^2 \times 5^2$, its square root also will have prime factors of 2, 3, and 5 but with the exponents halved. So the answer is $2^3 \times 3 \times 5 = 120$.

# Imperfect Squares

Most of the numbers that you'll encounter will not be perfect squares. A lot of whole numbers, like 12, are not perfect squares. And numbers whose decimal part is non-zero are never perfect squares. Depending on what a question asks for, your method of approach will vary.

If you're given a whole number and asked to find the square root and also are told that your answer should be "exact," then you'll use a slight variation on the method above. Whenever you're looking for an exact answer, don't be tempted to get out your calculator. If you use the calculator's answer—even if you copy all of the decimal places on the calculator screen—your answer will probably be wrong.

There are advantages and disadvantages to this situation. For one, it makes some problems so easy that you suspect that they are tricks. If you were asked to give an exact value for $\sqrt{2}$, the answer would be $\sqrt{2}$. That's right: there's nothing else you can do to that expression; the way it's stated originally is the exact value. On the downside, you will have to learn how to tell whether something is already in its simplest form or whether there is more work to do.

To find the exact value of the square root of a whole number, you'll need to start with it factored into primes and written in exponential form, then follow the methods described in the following examples.

**Example:** Find the exact value of $\sqrt{128}$, simplified.

**Solution:** This question is looking for an exact value, so don't even think of trying to answer with a decimal.

You start by factoring the entire number under the radical and replacing the number with its prime factorization. (In math-speak, we would call the number under the radical sign the "radicand." No one really talks that way except for math teachers.) So you have $\sqrt{128} = \sqrt{2^7}$. You can move numbers (and later, in algebra, letters) out from under the radical sign as long as you follow one simple rule: Two factors inside the radical sign become one factor outside the radical sign. Things have to move out from under the radical sign in factor pairs and the pair is replaced by a single copy of the factor. For example, because $\sqrt{2^7} = \sqrt{2 \times 2 \times 2 \times 2 \times 2 \times 2 \times 2}$, there are three pairs of 2s and one 2 leftover. Each time a pair moves out from under the radical sign, it becomes a single factor outside, so this simplifies to $2^3\sqrt{2}$. You might also choose to write this as $8\sqrt{2}$. Either form is correct. At least to me!

**Example:** Find the exact value of $\sqrt{600}$, simplified.

**Solution:** If you answered 24.5, you would get this wrong. If you answered 24.49489743, you would still be wrong. If you used the calculator built in to your computer and gave a gazillion decimal digits, you would be wrong again. This question is not looking for a decimal answer.

So what is it looking for? It's supposed to be an integer times something under a radical.

Factor 600, so $\sqrt{600} = \sqrt{2^3 \times 3 \times 5^2}$.

Look for pairs of prime factors. You have a pair of 2s, with one leftover, no pairs of 3s, and a pair of 5s. Fortunately you are not playing some sort of mathematical poker game, so it doesn't really matter how many pairs you have. Moving the paired factors out from under the radical sign (remembering that two factors inside become one factor outside), we get $2 \times 5 \times \sqrt{2 \times 3}$, which simplifies to $10\sqrt{6}$.

If you're looking for a decimal approximation, it's a different story. By far the simplest thing to do in that case is to use a calculator; within an instant you will have your answer. If you've been told that you need to estimate the value of a square root

**Timely Tips**

With radicals, the notation where there's nothing between the integer and the radicand for multiplication is the only way to go. Almost no one uses the × or · symbols to multiply a radical by an integer. Also, it's customary to have the integer on the left of the radical when multiplying them.

without a calculator, your best bet is to use guess and check. There is a method for finding square roots without a calculator and without using guess and check, but it has not been commonly taught in over 25 years; if you're curious about it, you can try to find a very old math teacher or else Google it.

Because you've memorized several whole numbers that are perfect squares, or at least have a cheat-sheet, you probably won't have to do any calculations to figure out which two whole numbers your answer is between. Once you have an initial guess, you can narrow down your search, trying smaller numbers if your answer is too big or larger numbers if your answer is too small.

**Example:** Without a calculator, estimate $\sqrt{70}$ to one decimal place.

**Solution:** Since 70 is between 64 and 81, $\sqrt{70}$ is between 8 and 9 (the square roots of 64 and 81).

Because 70 is closer to 64 than it is to 81, we'll guess that the answer is slightly closer to 8 than it is to 9. We'll start by guessing 8.2.

Getting out paper and pencil, $8.2^2$ is 67.24, which is a bit too small, so we'll try a larger number. Let's try 8.3.

Trying again, we calculate $8.3^2$, which is 68.89. We're getting closer! Next try: 8.4.

This time we calculate $8.4^2$ and get 70.56. That's really close!

**Practice Makes Perfect**

Problem 3: Find the exact value of $\sqrt{315}$, simplified.

Problem 4: Without a calculator, estimate $\sqrt{27}$ to one decimal place.

Because $8.3^2$ is less than 70 and $8.4^2$ is greater than 70, we know that the real answer has to lie between 8.3 and 8.4. Because $8.4^2$ is closer, when we give the answer to one decimal place, we say that it's 8.4.

At this point you can check with your calculator: $\sqrt{70} = 8.366600265...$, which rounds to 8.4.

## Dealing with Radicals

As we're not in medieval Europe, we can't deal with radicals by lopping off their heads. Instead we have to set strict rules on how to deal with them without their untimely demise. If you lost a radical in the middle of a problem you probably wouldn't be able to finish!

If you have two radicals, you can multiply them by taking the square root of their product. For example, $\sqrt{70} \times \sqrt{10} = \sqrt{70 \times 10} = \sqrt{700}$. Similarly, you can divide them by taking the square root of the quotient. For example, $\sqrt{70} \div \sqrt{10} = \sqrt{70 \div 10} = \sqrt{7}$.

It is very rare to be able to cleanly add or subtract two radicals. In other words, $\sqrt{70} - \sqrt{10}$ is *not* equal to $\sqrt{70-10} = \sqrt{60}$. Here's a good example of this:

You know that 9, 16, and 25 are perfect squares. $\sqrt{9} = 3$, $\sqrt{16} = 4$, and $\sqrt{25} = 5$. Thus it is fairly clear that $\sqrt{9} + \sqrt{16} = 3 + 4 = 7$.

However, $\sqrt{9+16} = \sqrt{25} = 5$. Since 5 and 7 are different numbers with different places on the number line, it is clear $\sqrt{9} + \sqrt{16}$ and $\sqrt{9+16}$ cannot be equal.

# Don't Be Square: Other Roots

Mathematicians love to generalize. That's a nice way of saying that once they figure out one trick or method to solve a problem, they then try to get as much mileage out of it as possible. You might call that "lazy," but in the math community it is considered "clever" and is called generalizing the solution. When it comes to square roots, there is no reason not to ask our basic question in a slightly different way. Instead of asking, "What number do you multiply by itself to get …?" you can ask, "What number do you multiply by itself three times to get …?" Or, "What number do you multiply by itself four times to get …?" And you can continue like this, giving you other options for roots besides square roots.

So what number do you multiply by itself three times to get 27? With a little checking, it's not hard to discover that $3 \times 3 \times 3 = 27$. The mathematical jargon for this idea is "cube root," and in math language, you would ask this question as, "What is the cube root of 27?" Again, there is also a notation for this. It looks fairly similar to the notation for square root, except there is a tiny 3 in the upper left to indicate that it's a cube root that you're looking for. In notation, we would ask $\sqrt[3]{27}$.

*A cube that is three units on a side is made up of 27 smaller cubes.*

There's no reason to stop at 3. Beyond this point, they no longer have special names and are just called fourth root, fifth root, and so on, and are denoted $\sqrt[4]{\phantom{x}}$, $\sqrt[5]{\phantom{x}}$, and such. The tiny little number in the upper left tells you which root it is. If we wanted to, we could write square roots as $\sqrt[2]{\phantom{x}}$, but almost no one ever does it that way.

With cube roots, to bring factors out from under the radical sign, three factors from inside the radical sign become one outside of it. For fourth roots, you'll need four copies of a factor to bring one out. And it continues like this for higher roots.

**Example:** Simplify $\sqrt[4]{144}$.

**Solution:** Start by factoring the number inside the radical sign into its prime factors. This gives you $\sqrt[4]{2 \times 2 \times 2 \times 2 \times 3 \times 3}$. Because this problem is about the fourth root, whenever you see four of a kind inside the radical sign, you can bring one copy out. Since there are four

> **Practice Makes Perfect**
>
> Problem 5: Find $\sqrt[3]{125}$, $\sqrt[4]{16}$, and $\sqrt[5]{16807}$.
>
> Problem 6: Simplify $\sqrt[3]{250}$, $\sqrt[4]{144}$, and $\sqrt[5]{\dfrac{69}{243}}$.

copies of 2, this becomes $2\sqrt[4]{3 \times 3}$. Since you don't have a full set of four 3s, they have to stay inside. This simplifies to $2\sqrt[4]{9}$.

# Rationalizing the Denominator

A lot of math teachers seem to have a phobia about fractions that have square roots (or any other roots) in the denominator. They just can't accept a fraction that looks like $\frac{1}{\sqrt{2}}$. It makes them nervous. Jittery. Upset. Unsettled. They just don't like it.

And you thought that you were the only one who had a complex about fractions. However, as your teacher is the one who assigns your grade in math class, you'll have to learn to accommodate this quirk and master the tricks for banishing square roots from the denominator of a fraction.

## A Single Square Root

The basic case is when there is one square root in the denominator. Something like $\frac{2}{\sqrt{5}}$ or $\frac{3}{5\sqrt{7}}$. It's okay if there are other numbers in the denominator, just as long as there is only one square root. In this case, you rationalize the denominator by multiplying both the numerator and the denominator of the fraction by the square root that appears in the denominator.

> **Timely Tips**
>
> Remember that multiplying both the numerator and the denominator of a fraction by the same number is the same as multiplying by 1. This keeps the value of the fraction unchanged.

**Example:** Rationalize the denominator of $\frac{7}{\sqrt{5}}$.

**Solution:** Multiply both the numerator and the denominator of the fraction by $\sqrt{5}$, giving you $\frac{7\sqrt{5}}{\sqrt{5} \cdot \sqrt{5}}$, which equals $\frac{7\sqrt{5}}{5}$.

**Example:** Rationalize the denominator of $\frac{2}{3\sqrt{6}}$.

**Solution:** You multiply both the numerator and the denominator by the square root that appears in the denominator, in this case by $\sqrt{6}$. This gives you $\frac{2\sqrt{6}}{3\sqrt{6} \cdot \sqrt{6}}$, which is the same as $\frac{2\sqrt{6}}{3 \cdot 6}$. Simplifying, you get $\frac{\sqrt{6}}{9}$.

---

Problem 7: Rationalize the denominator of $\dfrac{4}{\sqrt{3}}$.

Problem 8: Rationalize the denominator of $-\dfrac{12}{\sqrt{12}}$.

## Two Square Roots

If there are two square roots added together, the process is a little bit more complicated and it requires careful application of the Distributive Law. It's possible to rationalize denominators that are more complicated than this, but, most likely, no one will ask you to do it.

In order to do this type of problem, you need one piece of vocabulary and a little bit of practice with the Distributive Law. If you have two square roots added together, and you subtract them instead, you have what's called the *conjugate*. It is the same thing if you start with subtracting and then changing it to adding. It's still called the conjugate. For example, the conjugate of $\sqrt{12}+\sqrt{32}$ would be $\sqrt{12}-\sqrt{32}$. The conjugate of $\sqrt{17}-\sqrt{13}$ would be $\sqrt{17}+\sqrt{13}$.

What happens if you multiply the sum of two radicals by its conjugate? Well, if you are careful and use the Distributive Law accurately, all the square roots should go away.

**def•i•ni•tion**

The **conjugate** of the sum of a pair of roots is the difference of that pair of roots. The conjugate of the difference of a pair of roots is the sum of that pair of roots.

**Example:** Multiply $\left(\sqrt{8}+\sqrt{3}\right)\left(\sqrt{8}-\sqrt{3}\right)$.

**Solution:** Using the Distributive Law, this is going to be $\left(\sqrt{8}+\sqrt{3}\right)\left(\sqrt{8}\right)-\left(\sqrt{8}+\sqrt{3}\right)\left(\sqrt{3}\right)$.

Applying the Distributive Law again, you'll get $\sqrt{8}\cdot\sqrt{8}+\sqrt{3}\cdot\sqrt{8}-\left(\sqrt{8}\cdot\sqrt{3}+\sqrt{3}\cdot\sqrt{3}\right)$.

Next, distributing the minus sign, you'll have $\sqrt{8}\cdot\sqrt{8}+\sqrt{3}\cdot\sqrt{8}-\sqrt{8}\cdot\sqrt{3}-\sqrt{3}\cdot\sqrt{3}$.

Multiplying out the square roots gives you $8+\sqrt{24}-\sqrt{24}-3$, which is 8–3 = 5.

You might think that's a lot of work to go through just to get the number 5, and you'd be right. There is a short-cut way to multiply the sum of square roots by its conjugate: $\left(\sqrt{a}+\sqrt{b}\right)\left(\sqrt{a}-\sqrt{b}\right)=a-b$.

To prove this rule, you would do the exact same calculation as before, except you would write the letter *a* where all the 8s are and the letter *b* where all the 3s are. It's probably not worth your time to go through the calculation in detail; you'll be better off memorizing the rule.

To rationalize the denominator of something like $\dfrac{7}{\sqrt{2}+\sqrt{3}}$, you will multiply both the numerator and the denominator by the conjugate.

**Example:** Rationalize the denominator of $\dfrac{5}{\sqrt{11}+\sqrt{7}}$.

**Solution:** Start by multiplying both the numerator and the denominator by the conjugate of the denominator, giving you $\dfrac{5\left(\sqrt{11}-\sqrt{7}\right)}{\left(\sqrt{11}+\sqrt{7}\right)\left(\sqrt{11}-\sqrt{7}\right)}$.

You'll use the Distributive Law to multiply out the top, and the rule for multiplying conjugates to multiply out the bottom. This gives you $\dfrac{5\sqrt{11}-5\sqrt{7}}{11-7}$, which equals $\dfrac{5\sqrt{11}-5\sqrt{7}}{4}$.

Truth be told, this expression doesn't look much more pleasant than the one that was given at the beginning, but that isn't the point. The entire reason for doing this is to compensate for an irrational fear of radicals in the denominator of a fraction.

---

### Practice Makes Perfect

Problem 9: Multiply $\left(\sqrt{5}+\sqrt{3}\right)\left(\sqrt{5}-\sqrt{3}\right)$.

Problem 10: Rationalize the denominator of $\dfrac{5}{\left(\sqrt{17}+\sqrt{19}\right)}$.

---

# Rational Exponents

Have you wondered why the chapter on roots is in the exponents section? Exponents are written with small numbers in the upper right, while roots are written with a funky symbol surrounding the number, and when they do have a tiny number, it's in the upper left.

Exponents tend to make big numbers get even bigger; roots tend to make big numbers get smaller. Whatever one does, the other seems to undo. Ah, so there is a connection: roots and exponents go together in a pair. You can undo a root by applying an exponent, and you can undo an exponent by taking a root. For example, if you take the number 7 and square it, you get $7^2 = 49$. If you then take the square root of 49, you'll get $\sqrt{49} = 7$, restoring the number 7 that you started with. Exponents and roots are inverse operations.

There is another way that exponents and roots go together well. When you first learned about exponents, you learned how to raise a number to an integer power. You learned that raising to a positive exponent, like $11^3$, was really repeated multiplication, like $11 \times 11 \times 11$. And negative exponents were repeated division: $11^{-3} = \dfrac{1}{11 \times 11 \times 11}$. (Do you remember what happens if you raise something to the zeroth power? The answer is always 1.) Up to this point, we've been restricted to integers. Now we can extend the relationship between powers and roots by describing roots as fractional exponents.

## Simply Squares, Cubes, and Such

Rational exponents? Yes, you read that correctly, you can raise a number to the $\frac{1}{2}$ power. In fact, you can use any rational number as an exponent! The basic definition is that raising something to the $\frac{1}{n}$ power is the same as taking the $n$th root. So raising something to the $\frac{1}{2}$ power is the same as taking the square root; raising something to the $\frac{1}{3}$ power is the same as taking the cube root.

**Example:** Evaluate $64^{\frac{1}{2}}$.

**Solution:** Raising to the $\frac{1}{2}$ power is the same as taking the square root, so $64^{\frac{1}{2}} = \sqrt{64}$, which is just 8.

**Example:** Evaluate $27^{\frac{1}{3}}$.

**Solution:** $27^{\frac{1}{3}} = \sqrt[3]{27} = 3$.

> **Practice Makes Perfect**
>
> Problem 11: Evaluate $25^{1/2}$.
> Problem 12: Evaluate $216^{1/3}$.

## Powers and Roots

There are an awful lot of fractions out there that are not just the reciprocals of whole numbers. There are a lot of stupid-looking fractions like $\frac{7}{3}$ and the like, and all of them can be exponents. Being able to break these problems down and make them work relies on remembering the laws of math. You'll need to remember how fractions work, the Commutative Law, and the rules for raising something with an exponent to a still higher power. If you need to, go back and review Chapters 3, 4, and 7.

Simply put, if you're raising a number to a fractional power, you treat the numerator of the fraction as a power and the denominator of the fraction as a root.

**Example:** Evaluate $8^{7/3}$.

**Solution:** The numerator, 7, should be used as a power. The denominator, 3, should be used as a root—meaning you need to take the cube root. $8^7 = 2,097,152$, and $2,097,152^{\frac{1}{3}} = \sqrt[3]{2,097,152} = 128$.

Without a calculator, that example was fairly messy. An answer like 128 is fine to work with, but was there really a reason to go through an intermediate step that was larger than two million? Turns out that savvy application of the rules of math will allow you to take many problems of this type and make them so simple that you can do them without a calculator. Let's do the same problem again and take a closer look at the steps to minimize the amount of work that needs to be done.

**Example:** Evaluate $8^{7/3}$.

**Solution:** Working this problem in the most efficient way requires making a choice of the order to do things in. This choice comes from the fact that $7/3$ is, by the Commutative Law, both $7 \times 1/3$ and also $1/3 \times 7$. This, in itself, may seem like no big deal, but by the laws of exponents, this means that $8^{7/3}$ can be written as both $8^{7 \times 1/3} = (8^7)^{1/3}$ and also as $8^{1/3 \times 7} = (8^{1/3})^7$. The first of these was how this example was done the first time—yielding an intermediate step involving a number over two million. This time, try evaluating the second version. $8^{7/3} = 8^{1/3 \times 7} = (8^{1/3})^7 = 2^7 = 128$. Much easier.

---

**Practice Makes Perfect**

Problem 13: Evaluate $25^{3/2}$.

Problem 14: Evaluate $64^{4/3}$.

---

# Negative Rational Exponents

These are the most complicated exponential expressions that you will have to deal with for a long, long time—possibly the most complicated ones that you will ever see.

Making sense of these problems comes down to breaking them down into pieces:

◆ The numerator tells you the power.

◆ The denominator tells you the root.

◆ A negative sign tells you to take the reciprocal.

**Example:** Evaluate $27^{-2/3}$.

**Solution:** There are a bunch of things to look at in this problem. The 2 in the numerator of the exponent tells you that you'll be squaring. The 3 in the exponent's denominator says that you'll be taking a cube root. And the fact that the exponent is negative means that you'll need to take a reciprocal.

The great thing about this type of problem is that the commutative property of multiplication and the rules of exponents promise you that the order doesn't matter when you do these steps! Here is your payback for the times you were tripped up by the order of operations. In problems like this, the steps don't matter.

More often than not, it is easier to take the root first, and that applies in this case, too. So your first step is to take the cube root of 27, which is 3.

Now, 3 is not your answer; you still need to square it and take its reciprocal. Squaring, you'll get 9; take the reciprocal, you'll get $\frac{1}{9}$. So $27^{-2/3} = \frac{1}{9}$.

**Practice Makes Perfect**

Problem 15: Evaluate $25^{-3/2}$.

Problem 16: Evaluate $64^{-4/3}$.

So far we have defined whole number exponents as repeated multiplication, negative exponents as repeated division, and now rational exponents as powers of roots. Until you reach the point of calculus, every exponent you see will probably be a combination of these types. In fact, any number can be an exponent, but defining what that means (and using it) is typically reserved for higher math.

## Let Our Powers Combine

We now know what to do with any rational exponent. But what happens when we have multiple exponents over the same number? This can arise in two situations. The first is when the base and the exponent are raised to some other rational exponent, like $(27^{-2/3})^2$. In this case, multiply the outer and inner exponents and write the product as the exponent of the same base. The second case is where the exponent itself is raised to some power, such as $2^{9^{\frac{1}{2}}}$. In this case, evaluate the entire expression in the exponent first. The basic rule for evaluating exponents of exponents of exponents of … is that you must evaluate the highest-level exponent first and work your way down.

**Example:** Rewrite $(27^{-2/3})^2$ as an expression with only one exponent.

**Solution:** Because the base-exponent pair is raised to the power of 2, we multiply the two exponents and put them over the original base. $(27^{-2/3})^2 = 27^{-2/3 \times 2} = 27^{-4/3}$.

**Example:** Rewrite $2^{9^{\frac{1}{2}}}$ as an expression with only one exponent.

**Solution:** Because the exponent itself is raised to the power of $\frac{1}{2}$, you need to start by thinking about what that means. Raising to the $\frac{1}{2}$ power is the same thing as taking the square root; there's nothing to understand about that, you just need to accept it and remember it. Raise 9 to the $\frac{1}{2}$ (which is to say, take the square root of 9)

and write the answer as the exponent for the original base. $2^{9^{\frac{1}{2}}} = 2^{\left(9^{\frac{1}{2}}\right)} = 2^3$

# Using Exponents to Simplify Fractions

You can use your newfound knowledge of exponents and the rules for multiplying and dividing them to help when simplifying fractions.

**Timely Tips** _____

In Chapter 7 you learned how to use exponential notation to express a whole number as a product of primes in a compact way. You'll be making use of that when simplifying fractions in this section. If you've forgotten how to factor whole numbers, see Chapter 1.

## Answering in Fraction Form

One of the beautiful interplays in mathematics is how well fractions and exponents play with each other. If you wish, you never have to use a negative exponent! Alternatively, if you despise fractions you can live without them by using exponents in their place. You'll need to be able to work with one or the other: fractions or negative exponents. If you don't, you can't do as much and everything starts breaking down.

To annihilate negative exponents, take the reciprocal of your base and raise it to the negative of your old exponent.

**Example:** Eliminate all negative exponents from $9^{-7}2^3 17^{-1}5^0$.

**Solution:** The 0 exponent over the 5 takes care of that term nicely since $5^0 = 1$.

Both 9 and 17 have negative exponents, so in each case we should replace them with the reciprocal of the base raised to the negative of the current power.

This isn't as confusing as it sounds. Remember that negative exponents can be interpreted as dividing by the base multiple times? We're simply rewriting the negative exponent division problem as a fractional division problem. The reciprocal of 9 is $\frac{1}{9}$, and the negative of –7 is 7. This means we should replace $9^{-7}$ with $(\frac{1}{9})^7$.

Similarly, we transform $17^{-1}$ into $(\frac{1}{17})^1$. We'll drop the exponent of 1 since raising a number to the first power equals itself, just like adding 0 or multiplying by 1 gives the original number back to you. Though my desk wouldn't believe I think this, getting rid of unnecessary clutter almost always helps!

Returning to the original equation and substituting in the fractions we constructed, we get $\left(\dfrac{1}{9}\right)^{7} 2^{3}\left(\dfrac{1}{17}\right)$. It's customary to replace $2^3$ with $\dfrac{2^3}{1}$ and multiply the fractions together to get $\dfrac{2^3}{9} \times 17$.

> ### Practice Makes Perfect
>
> Problem 17: Rewrite $\left(8^{-\frac{5}{2}}\right)^{64^{\frac{5}{6}}}$ as an expression with only one exponent.

# Using Positive and Negative Exponents

Now it's time to rid ourselves of fractions! We're almost at the end of the chapter, so bear with me for just a little longer. We're going to reverse the procedure from the previous practice problems.

1. Start with a fraction.

2. Move the denominator from below the numerator.

3. Raise the denominator to –1.

4. Multiply by the numerator.

**Example:** Change $\dfrac{4^5}{23^2} \times 7^3$ from fraction notation to exponential notation.

**Solution:** The denominator of $\dfrac{4^5}{23^2} \times 7^3$ is $23^2 7^3$. Let's pull out the denominator from beneath the numerator and place it somewhere else, leaving us with $\dfrac{4^5}{1}$. Raising the denominator to –1, we get $(23^2 7^3)^{-1}$.

Multiplying this by what was left over in the fraction, just the numerator, gives us $4^5(23^2 7^3)^{-1}$.

You should always check your work if you have time, and this type of problem is easier to check than most.

Start with $4^5(23^2 7^3)^{-1}$. An exponent of –1 means we take the reciprocal of the base, which in fractional form is $\dfrac{1}{23^2} \times 7^3$.

---

### Practice Makes Perfect

Problem 18: Change $\dfrac{4^{-2}3^7}{5^{-1}7^4 11^3}$ from fraction notation to exponential notation.

Problem 19: Change $2^4 3^{-5} 21^{13} 9^{-7}$ from exponential notation to fraction notation.

Substitute the negative exponent and its base with this fraction and make $4^5$ into a fraction by putting it over 1.

Our expression now looks like this: $\dfrac{4^5}{1} \times \dfrac{1}{23^2} \times 7^3$.

Performing multiplication of fractions,

$$4^5 \times \frac{1}{1} \times 23^2 \times 7^3 = \frac{4^5}{23^2} \times 7^3,$$ we end up with

precisely what we originally started with.

Changing fractions into positive and negative exponents sometimes helps you see how to solve problems. By looking at the problem in several different forms, you may understand it better and have insight you may otherwise have missed.

## The Least You Need to Know

- ◆ Square roots ask what number you have to multiply by itself to get the desired number.

- ◆ When taking square roots, a pair of factors under the radical sign becomes a single factor outside it.

- ◆ The denominator of a fractional exponent tells you which root to take.

- ◆ The numerator of a fractional exponent tells you which power to raise things to.

# Chapter 9

# Decimals, Addition, and Subtraction Meet Exponents

## In This Chapter

◆ Base ten and powers of ten

◆ Scientific notation and significant figures

◆ Financial exponents

◆ Mathemagical connections in math

You've almost made it! You may be hefting this book, looking at the pages elapsed and the pages still to go, and thinking that the end looks anything but near. But, in some sense, it is. In the previous chapters you have worked with all sorts of numbers as well as an assortment of ways of mixing and combining them. In this chapter you'll learn about a few critical applications and properties of exponents and roots.

Most of the applications of square roots (and higher roots) are in problems where something has been multiplied by itself and you have to undo the process. In algebra class this will manifest itself in the dreaded quadratic equation and solving problems where you have to figure out when a rock, dropped off the edge of a cliff, hits the ground. However, in pre-algebra, the applications of roots and exponents are much more modest.

# Base Ten

In 1202, a merchant's son named Leonardo Pisano of the Family Bonacci (better known by his nickname Fibonacci) published a book called *Liber Abaci*, or the *Book of Calculation*. His book covers most of the same mathematical topics that are included in this one: working with integers, calculating with fractions, solving problems involving proportions. His goal was to teach the Italian merchants how to do basic math problems related to commerce.

One key difference between his book and this one is that Fibonacci needed to begin his book with a description of the base-ten number system we take for granted. (There is a refresher on base-ten numbers in the next few paragraphs just in case.) At the time he was writing, most of Europe was still using Roman numerals for all their calculations. We might be okay with expressing the copyright date for a film as MMVII, but can you imagine calculating CXCVI × MLIV? Or even IX + II? Even if you complain about having to calculate things by hand, you should be grateful that you are using a base-ten number system and not Roman numerals.

So what is a base-ten number system? You might have faint memories from elementary school of being told to use base-ten blocks and to collect units into groups of ten called longs and then combining ten longs into a group of 100 called a flat. The key idea is to take your quantity and group it into powers of ten and then report how many you have of each size group.

If you're dealing with whole numbers, your groups will have $10^0 = 1$, $10^1 = 10$, $10^2 = 100$, $10^3 = 1,000$, and so on. You need to group things as efficiently as possible, so you group things in order to have fewer copies of a larger group instead of many copies of a smaller group.

**Timely Tips**

The arithmetic of computers is done with only 0s and 1s. This is the base-2 number system, in which each place value is a power of 2. Sometimes computer scientists find it useful to use a base-16 number system called hexadecimal. Hexadecimal notation uses the digits 0-9 together with six additional digits called A-F, and each place value represents a power of 16.

Let's imagine that you had a one-pound bag of M&Ms and you were asked to represent the number of M&Ms in the bag as a base-ten number. In the real world you would just count them using any efficient method that you liked and report the answer. Unless you are a computer genius or have some sort of binary hobby, every number that you deal with is a number in base ten.

However, this is a math problem, so you'll have to do things a little bit differently. First you divide the M&Ms into groups of ten until you can't do that anymore. At this point, you'll have a bunch of piles of ten M&Ms and one smaller pile with fewer than ten. Don't eat those yet!

Now what you'll do is take your piles of ten and combine ten of them into piles of 100 M&Ms. At this point, you'll have some piles of 100 M&Ms, a few leftover piles of ten M&Ms that couldn't be combined, and the little pile.

Are there more than 1,000 M&Ms in a one-pound bag? I'll admit that I haven't actually counted before creating this example because I am a theoretical mathematician and don't usually bother with checking how things work in the real world. In my defense, I suspect that none of the readers of this book are currently counting all the M&Ms in a one-pound bag, either. If there are more than 1,000 M&Ms in a bag, you can take ten of your piles of 100 and combine them into a pile of 1,000. You'll keep doing this until you have consolidated your M&Ms efficiently into piles whose sizes are powers of ten.

To represent this quantity as a base-ten number, you count how many piles you have of each size and write this digit down. Suppose you have five piles of 100 M&Ms, one pile with ten, and then seven leftover M&Ms off to the side. The base-ten representation of the number of M&Ms is 517. If you had five piles of 100, no piles of ten, and four M&Ms off to the side, then you'd write this in base-ten as 504.

Now that you've had your grubby hands all over the M&Ms while sorting and counting them, you can eat them all. At this point nobody else is going to want them.

# Scientific Notation

A fair amount of science involves measuring things, collecting data, and making calculations. Mathematics is the language of science, and many scientific ideas and measurements are described in terms of numbers. Scientific notation expresses a number as a number between 0 and 10 multiplied by a power of ten. A number in scientific notation will look something like $6.02 \times 10^{23}$.

**Timely Tips**

Because calculator screens can't show exponents as tiny superscript numbers, many of them use the notation E to mean "times ten to the power of...." To a calculator, the number $6.02 \times 10^{23}$ would usually be written as 6.02E23.

# Significant Figures

Every measurement has a certain amount of inaccuracy, and scientists want to be able to communicate how accurate a number is. Do you know exactly how many ants live in your ant farm, down to the last individual? (If you've given them names, you should either consider a career in entomology or else find a new hobby.) Or do you only know to the nearest 1,000 ants? In scientific notation, only the digits that are certain are given. If we're certain about a digit, we call it a *significant digit*, and scientists often talk about the total *number of significant digits* in a number. Scientists also want to be able to easily compare the sizes of different numbers. With the numbers we see in everyday life, that's not such a big deal. However, when it comes to very large numbers with a lot of trailing zeros, you'd have to count them all to be sure how big the number was.

> **Kositsky's Cautions**
>
> Don't think that you can get out of scientific notation by using words like "billion" to describe the size of large numbers. Americans use the word billion to represent 1,000,000,000, but people in the United Kingdom call 1,000,000,000,000 a billion!

# Very Big Numbers

The distance to the moon is 384,400,000 meters. It's not exactly 384,400,000 meters; it could be a bit more or less. However, the figure 384,400,000 is accurate to the nearest 100,000 meters.

Converting the number to scientific notation is relatively straightforward. First, locate the spot between the first digit and the second digit; in this case, that's between the 3 and the 8. Now you count how many digits there are to the right of that spot. This number will give you the exponent for the 10. Looking at the distance to the moon, you can count that there are eight digits to the right of the 8. Finally, you put a decimal point between the first and second digit, drop any trailing zeros, and multiply by 10 to the power that you counted before: $3.844 \times 10^8$ meters.

**Example:** The mass of Earth is 5,974,200,000,000,000,000,000,000 kg. Write this in scientific notation.

**Solution:** We start counting at the spot between the 5 and the 9, and see that there are 24 digits to the right of that spot, so the exponent on 10 will be 24. In scientific notation, the mass of Earth is $5.9742 \times 10^{24}$ kg.

Sometimes we want to take a number in scientific notation and convert it back into *standard notation*. Standard notation is just a fancy name for the way that we usually write numbers. When you're looking at a number like 23847 or 45379238492347203 or any other number written out in the usual way, it's written in standard notation. When dealing with large numbers, it's fairly simple. You just multiply out the expression, remembering that multiplying by 10 raised to a positive number means that you move the decimal point that many places to the right; once you reach the right-hand side of the number, you can keep moving the decimal point to the right by adding more zeros to the right of the number.

## def•i•ni•tion

When we talk about **standard notation**, we mean the normal way of writing numbers that we use all the time. It's nothing special.

**Example:** The planet Neptune is about $4.3 \times 10^{12}$ meters from Earth. Express this distance in standard notation.

**Solution:** Start with the 4.3 and move the decimal point 12 places to the right. Now, after you've moved it one space to the right, you're at the edge of the number, so how do you continue to move it? Simple. Each time you need to move another place to the right, you tack another zero on to the right edge of the number. This gives you 4,300,000,000,000 meters. This is the same answer that you would have gotten if you had carried out the multiplication in the original problem; it's two ways of saying the same thing.

---

### Practice Makes Perfect

Problem 1: Marvin has 2,345,000,000,000,000,000,000 hamsters living with him. Write the number in scientific notation.

Problem 2: Avogadro's number is $6.02 \times 10^{23}$. Write the number in standard notation.

---

## Very Small Numbers

Not all of science studies large numbers. There's more to consider than the distance to faraway worlds and the number of bacteria in a dish. Sometimes we need to represent tiny decimals, numbers very, very, very close to zero. Scientific notation can make these numbers clearer to deal with, too. Instead of using positive exponents on the power of 10, we represent decimals in scientific notation by using negative powers of 10.

The process of converting a very small number to scientific notation is similar to— but not the same as—the method used for very large numbers. When working with decimals that are less than one, you need to locate the left-most non-zero number to the right of the decimal point. That's a bit of a mouthful, and it's easier to find that place than to describe how to find it.

Let's work with an example: the width of a dust mite is about 0.000125 meters, and you want to express this in scientific notation. Here you're looking at the spot between the 1 and the 2. Now you need to count how many places there are between this spot and the decimal point. In this example, the spot between the 1 and the 2 is four places to the right of the decimal point. The number of places that you count (in this example, four) will be the exponent on the 10—with the important note that the exponent must be negative.

To finish the problem off, you just need to put the decimal point after the left-most non-zero number, include all the rest of the digits, then multiply that by 10 to the number that you counted. In scientific notation, the width of the dust mite would be $1.25 \times 10^{-4}$ meters.

**Example:** A beam of red light has a wavelength of 0.0000635 meters. Express the wavelength in scientific notation.

**Solution:** The distance between the decimal point and the spot between the 6 and the 3 is 5 places, so the exponent on 10 will be –5. Therefore, in scientific notation, this number is $6.35 \times 10^{-5}$ meters.

Just as with large numbers, converting back to standard notation requires nothing more than multiplying out the expression. Where with large numbers we moved the decimal point to the right, when we have tiny decimals, the negative exponent tells us to move the decimal point to the left. If you're moving more than one place to the left, you'll have to add extra zeros at the beginning of the number to make that possible.

**Example:** The width of a piece of spider silk is $5 \times 10^{-6}$ meters. Express this number in standard notation.

**Solution:** You need to move the decimal point six places to the left, giving you 0.000005 meters.

| Practice Makes Perfect |
|---|

Problem 3: The gravitational constant G is $6.67 \times 10^{-11}$. Write the number in standard notation.

Problem 4: The mass of a neutron is 0.00000000000000000000000167 grams. Write the number in scientific notation.

# Financial Math

You've probably heard the stories of celebrities who have ended up destitute because their managers cheated them out of a lot of money. While many of these celebrities may have been too busy to take care of their finances, some of them probably abdicated the responsibility because they didn't see themselves as numbers people. If you don't have a decent understanding of the basics of math, you won't be in a position to make good financial decisions for yourself.

The most important mathematical idea in financial math is exponents. Most calculations involve taking 1 plus the interest rate and raising it to a power, then multiplying this by an amount of money. This shows you how your money can grow when you invest it. It also makes it so that you owe more and more money if you stay in debt.

Figuring out the balance of a savings account or of a debt is based on one principle. If you are letting someone use your money, you want something in return. And that is more money. Even though more complicated investments work differently than savings accounts in financial markets, the calculation to figure out how much is in your account is the same.

## Investment

Interest, either interest earned on a savings account or interest paid on a debt, is calculated as a percentage. If you have an interest rate of 5% a year, this means that for every $100 in the account, the interest will be $5 a year. If it's your money in a savings account, the bank will pay you. If you've borrowed money, you pay that $5 to the bank. If your account balance is something other than $100, the way to calculate this in general is to take the bank balance and multiply it by the interest rate written as a decimal.

**Example:** You have a balance of $260 at an interest rate of 4 percent. What is the interest?

**Solution:** (260)(0.04) = 10.4, which is $10.40.

To find the new balance in the account, you add the interest to the balance. In the example above, the balance will be $260 + $10.40 = $270.40. The way this was calculated turned it

---

**Practice Makes Perfect**

Problem 5: You've been hired by the United Wildebeest Company and they offer you two possible signing bonuses:

1. $10,000 in an account that earns 10% interest per year.

2. $8,000 in an account that earns 15 percent interest per year.

If you don't need the money for 10 years, which should you choose?

into a two-step process: first you calculated the interest and then you added it to the old balance.

If the question doesn't ask you to calculate the interest separately, you can calculate the new balance in just one step. Take the old balance and multiply it by 1 plus the interest rate (written as a decimal).

**Example:** You have a balance of $930 at an interest rate of 3%. What is the balance after one year?

**Solution:** $930(1 + 0.03) = 957.9$, which is $957.90.

When you invest over a number of years, you'll earn interest on your interest. In that case, you use exponentials to calculate the investment balance.

If you invest a principal $P$ at an interest rate $r$ for $Y$ years, the balance in the account is $P(1 + r)^Y$.

**Example:** You invest $730 in an account earning 4.5% for 12 years. What is the account balance?

**Solution:** $730(1 + 0.045)^{12} = \$1237.99$. Remember to always round your answer to the nearest cent.

# Endowments

Harvard University has a lot of money to spend. Harvard is a very, very wealthy organization, even wealthier than some countries! At the time that this book was written, its endowment, which grows each year, was about $35 billion. That's $35,000,000,000. You might want the chance to spend some of that $35 billion, but Harvard doesn't.

They don't spend the endowment, they try to make it earn more money in investment interest. Harvard has a special team of fund managers that tries to get the endowment to earn as much money as possible. If the endowment earns 5 percent interest in a year, that would be $(35,000,000,000)(0.05) = \$1,750,000,000$.

Maybe you can't compete with Harvard but you can still benefit from the idea. Let's imagine that your grandmother wants to put a provision in her will so that a fund will provide for spending money for your

---

| Practice Makes Perfect |
| --- |
| Problem 6: Your parents set up a $1,500,000 endowment to pay your living expenses. If it's earning an interest rate of 3.5 percent, how much will you have to live on per month? Remember, you're not allowed to touch the endowment! |

uncle who can't hold down a job. Maybe she thinks that Uncle Rob should get $300 a month to cover his text messaging, gas money, and trips to the arcade. (There's probably a reason that he can't hold down a job.)

If he's going to get $300 a month, the account is going to need to generate 12 times that much in interest each year, or $3,600 a year. So now the question is: How much should she set aside in her will so that the account can earn $3,600 in interest each and every year? She's going to have to make one assumption: she's going to have to guess at the interest rate. Let's say that your grandmother decides that a 3.5 percent interest rate is a safe assumption.

To calculate the amount she needs to fund the account, she'll take the yearly income and divide it by the interest rate, $3,600 \div 0.035 = \$102,857.14$. Uncle Rob had better hope that Grandma has a lot of money so that she can set up his account. Otherwise he might need to put in an application at the coffee shop.

## Mortgages

When you buy a house, you get a type of loan called a mortgage. The first syllable of the word, pronounced "more," is spelled "mort," a Latin word for death. This is because this type of loan is meant to kill the debt. A mortgage payment is calculated so that each month the payment is equal (all payments are the same number of dollars), every payment includes the fair share of interest owed to the lender, and the total payments will exactly pay off the debt in the agreed-upon period.

Calculating a mortgage payment requires a few pieces of information:

- The amount borrowed, called the principal, and abbreviated $P$.

- The annual interest rate, $r$, written as a decimal.

- The number of years over which the money will be paid back, abbreviated as $Y$.

The formula for calculating the mortgage payment $PMT$ is:

$$PMT = \frac{P \times \left(\dfrac{r}{12}\right)}{1 - \left(1 + \dfrac{r}{12}\right)^{-12Y}}$$

This formula has it all. It has decimals and fractions. It has order of operations. It has negative exponents. If you can calculate a mortgage payment, you have mastered the basic operations of pre-algebra. A lot of my students are able to set up the problem but have trouble putting the information into their calculators.

**Example:** You borrow $103,200 from the bank at an interest rate of 5.25% and agree to pay it back over a period of 15 years. What is your monthly payment?

**Solution:** Your payment is $\dfrac{103,200\left(\dfrac{0.0525}{12}\right)}{\left[1-\left(1+\dfrac{0.0525}{12}\right)^{-(12\times15)}\right]}$ . This represents a number, but it's

not the sort of number that's especially useful to most people in everyday situations. You wouldn't put that large expression in your monthly budget.

The best way to deal with simplifying this is in stages and to start with the denominator. The product in the exponent is a good place to start: $12 \times 15 = 180$.

It's hard to write this out step by step on paper because to keep the answer accurate, you want to keep the numbers in your calculator memory. There are more digits of accuracy than show up on the screen. If you write down the numbers that appear on your calculator after each step, you'll make rounding errors.

You'd probably start by calculating $0.0525/12$, adding 1 to that, and then raising the number in your calculator to the −180 power.

From there, you would use the memory feature to subtract that from 1. Press the reciprocal button (marked $\frac{1}{x}$), multiply by 103,200, multiply by 0.0525, and divide by 12. You should get $829.60.

That's the curse of the calculator. It makes each step easier, but the entire problem becomes much more complicated.

# Mathemagical Connections

One of the magical parts of mathematics is how everything can be used for a purpose other than its original intent. Math as a field is kind of like a county or a state where everyone is related to everyone else. No matter where you look, you're surrounded by relatives. On the plus side, mathematical relatives are often useful and interesting!

## Areas of Squares

When you first learned about multiplication, you may have drawn diagrams that are called *arrays*. To multiply 5 × 3, you may have drawn a box that was five units in length and three units in width, and counted how many squares were in the box. The total number of squares was the side length multiplied by the width. If your box is a square, then the number of boxes it's made up of is the side length times itself, or squared. This is how *squaring a number* got its name. If you want, you can make squares out of the M&Ms from earlier in the chapter. They're much more tasty than boxes.

The area of a square is defined to be the number of 1 × 1 squares inside it, including parts of 1 × 1 squares. These problems can be worked in two ways. If you're told the length of the side of the square, you can find the area by squaring the length of the side. However, if you're told the area of the square, you can rediscover the length of the side by taking the square root of the area. The units on side lengths are always raised to the first power, and units on area are always raised to the second power. "Square meters" really means "meters squared," or the equivalent of some number of 1 meter by 1 meter squares.

## def•i•ni•tion

An array is way of arranging things in a rectangle so that they're lined up in rows and columns. The desks in many classrooms form an array.

*Square meters.*

| a | $\sqrt{S}$ | 1 | 2 | 3 | 4 | 5 | 6 | 7 | 8 | 9 |
|---|---|---|---|---|---|---|---|---|---|---|---|
| $a^2$ | S | 1 | 4 | 9 | 16 | 25 | 36 | 49 | 64 | 81 |

...

**Example:** A square has a side length of 5 meters. What is the area of the square?

**Solution:** $5^2 = 25$, so the square has an area of 25 square meters.

**Example:** A square has an area of 36 square meters. What is the length of one of its sides?

**Solution:** $\sqrt{36} = 6$, so the side of the square is 6 meters long.

### Practice Makes Perfect

Problem 8: The side of a square has a length of 2.5 meters. What is the area of the square?

Problem 9: A square has an area of 49 square meters. What is the length of one of its sides?

## Solving Problems with Primes

Knowing the square root of a number also can help you in the process of searching for its prime factors. There aren't any really good ways of checking to find out which primes divide into a number. You are going to be stuck checking possible factors by hand. The luckiest that you're going to get with this type of problem is if you're allowed to use a calculator. Aside from that there are no tricks or special shortcuts that will make your life any easier.

However, there is one thing that you can use to your advantage. If you know what the square root of the number is, then you know when you can stop checking. The factors of a number come in pairs. One member of the pair is always less than or equal to the square root; the other member of the pair is always greater than or equal to the square root. There is one class of exceptions to this rule. When a number is a perfect square, its square root doesn't have a mate. What all this means for you is that when you are searching for factors, you can search up to the square root (including the square root itself if it's a whole number). Once you've checked all of those numbers, you can stop.

**Example:** What numbers do you need to check to find all of the prime factors of 700?

**Solution:** With a calculator, you can find that $\sqrt{700}$ is a little bit bigger than 26, so you need to check the primes up to 26. Once you've tested to see if any of the primes 2, 3, 5, 7, 11, 13, 17, 19, and 23 divide into 700, you're done!

---

### Practice Makes Perfect

Problem 10: What numbers do you need to check to find all of the prime factors of 802?

Problem 11: Find all of the prime factors of 782.

---

## A Magical Mystery Tour of Numbers

You've probably noticed that many problems involving zero and one are pretty easy. Adding and multiplying with zero is really easy. Multiplying and dividing with one is easy, too. Because these numbers are so nice to multiply, they also have some special properties when working with exponents.

◆ 1 raised to any power is still 1. It doesn't matter what the exponent is. If you take 1 and raise it to any number that you can think of, the answer is still 1.

◆ Any number raised to the zeroth power will be 1.

◆ Raise any number to the first power, and you get the number itself back as an answer. This is like multiplying by 1 or adding zero.

◆ Having an exponent of –1 means that you take the reciprocal.

These are four of the simplest, most useful properties of exponents. Although they may seem confusing at first, memorizing them will help you quickly solve a variety of problems.

## The Least You Need to Know

◆ Scientific notation expresses a very large or a very small number as a number between 0 and 10 times the appropriate power of 10.

◆ The number of significant figures increases with the quality of a measurement.

◆ Financial wealth can grow exponentially.

◆ If you know the area of a square, you can calculate the length of its sides by taking the square root of the area.

# Part 4

# Self-Expression Through Algebra

In cartoons and comic strips, jumbles of letters, numbers, and symbols litter chalkboards as a sign to the reader that something arcane and complicated is going on or that one of the characters is really smart. There's a method to our madness, and what looks like a mishmash of nonsense really makes sense if you know the rules. In this part, we let you in on the secret of how to do math with letters, and you'll be able to create your own impressive symbolic masterpieces.

*"Why can't you just draw antelope like everyone else?"*

# Chapter 10

# It's Just an Expression

## In This Chapter

- ◆ Math with letters
- ◆ Notations for working with variables
- ◆ Distributive Law and FOIL
- ◆ Plugging in numbers for variables

When you see math depicted in cartoons, there is often a classroom blackboard covered not with numbers, but with a gibberish of letters held together by mathematical symbols. There's some sort of cultural expectation that the secrets of math have nothing to do with numbers, rather they are hidden in an arcane language of nonsense letters like $x$.

In this chapter, you start learning how to translate the mathematical language you already know into these more arcane forms and you find out how to use this mathematical dialect (i.e., gibberish) to describe quantitative relationships from everyday life.

# Where Have the Numbers Gone?

Up to this point, just about everything you have seen in a math problem has been a mathematical question waiting for you to solve it by following a straightforward series of steps. Even the meanest and nastiest order of operations problem can be solved to get the right answer by carefully following the rules.

Now we're going to introduce expressions and variables. These are mathematical objects that like to stand around looking menacing. On their own they aren't really asking a question, but they can be used as part of math problems, like $x + y$. What do you do with $x + y$, you ask? Right now, nothing. There is no answer. You can't add it and you can't enter it into your calculator for an answer. You need to keep reading.

## Variable Names

A *variable* is something that stands in for a number when we don't know which number we're talking about. Just like "hey you" or "that guy" has only limited effectiveness when talking to or about people, it's equally awkward to keep trying to explain something in terms of "that first number that you multiplied by before you added it to that other number." In order to deal with this type of situation, we tend to give names to things. Now we might call that first number you multiplied by before you added to that other number something more memorable, like Rufus. More likely, instead of giving it a name like a person, we'll just give it a single letter name, like $x$.

Let's say you had two numbers—it doesn't matter what they are—and you decided to call them $x$ and $y$. This is really the essence of what's going on here. Sometimes you'll be talking about a mathematical idea, and it really won't matter what the numbers are. When you're talking about a property that holds for all numbers, it doesn't matter which numbers you're talking about.

**def•i•ni•tion**

A **variable** is a symbol that is a placeholder for a number. Most variables you see will be Greek or Latin letters. Unless specified otherwise, variables can take the value of any real number.

Sometimes, however, you'll be talking about a number in the abstract, calling it by some name like $x$, and it really does have an actual, specific value. You just don't know what it is yet. Any time it's inconvenient, awkward, or inappropriate to talk about a specific number, we tend to give it a name.

Back to the situation with your two numbers $x$ and $y$: to describe the process of adding the first number to the second number, you could just say $x + y$. If you wanted to subtract 2 from your first number, you'd say $x - 2$.

## Naming Traditions

All this said, you can use pretty much any letter from any alphabet as a variable in a mathematical expression. It's traditional to use letters from the regular alphabet and sometimes from the Greek alphabet, but nothing is stopping you from using letters from Russian, Hebrew, Arabic, Chinese, or any other script. Here are some of the basic conventions in use in many mathematical contexts.

- Letters from the end of the alphabet, like $x$, $y$, and $z$, are most frequently used as variables. They're especially common when your problem is very abstract, where it doesn't matter what the variable stands for, when you're considering a lot of cases at once, or when the person writing the problem isn't very creative.

- Letters from the beginning of the alphabet, like $a$, $b$, and $c$, typically

**Amy's Answers**

Because $x$ is such a common variable name, an often-used notation for multiplication is a dot, like this: $3 \cdot 5$.

are used when stating properties. For example, we could describe the Commutative Law for addition as $a + b = b + a$. We don't know what numbers $a$ and $b$ stand for, but it doesn't matter. Throughout the problem, $a$ stands for one number and $b$ stands for another, and it doesn't matter which ones because the Commutative Law applies to all numbers. Letters at the beginning of the alphabet often are used for numbers whose values we don't know at the moment but that once we pick them would be fixed.

> **Kositsky's Cautions**
>
> At the point where letters and numbers start to jumble together in your math homework, you'll need to take a close look at your handwriting. Is that a *2* or a *z*? A *t* or a *+*? Have you written a *5* or an *s*? An *x* or a *×*? Neat handwriting may be the difference between the right answer and the wrong one.

♦ Letters that begin words in the problem normally represent those words. If you're working on a problem that involves a bunch of objects like trees and cars and deer, you might let $t$ stand for the number of trees, $c$ for the number of cars, and $d$ for the number of deer. Unless someone tells you otherwise, however, the variable $t$ usually stands for time.

$$xxlll = SS$$

♦ Some letters have special meanings. Mathematicians have special meanings for some letters, so they don't show up all that often as variables. The letter $e$ stands for a number close to 2.718 and that is used a lot in calculus. The letter $i$ represents a specific imaginary number. You won't see the letter $o$ used very often because it looks a lot like a zero.

# Mathematical Expressions

Just about any combination of letters and numbers held together with mathematical symbols and written in a way that is consistent with mathematical notation could be called an *expression*. This is a bit of a catch-all term, and can be used to describe things made up of only numbers and operations as we have done in the first half of this book, or it could be used for more complicated constructions as well. More often than not, however, when we use the term expression we are talking about something made up of numbers, letters, and symbols, but which does not include an equals sign. Once you have an equals sign involved it becomes an equation, which is a matter for the next chapter.

## def•i•ni•tion

An **expression** is a jumble of numbers, letters and symbols that takes on a numerical value when all variables are assigned values.

## Operations

The same operations you're familiar with from numbers also will work when you include variables in an expression. You can add, subtract, multiply, and divide. Fractions can include expressions. You can raise numbers to powers and extract roots. Anything that you did with numbers you can, and will, do with expressions.

The one big difference you'll see when working with expressions is that the notation for multiplication and division will change. I'm not really sure how this got started or why this is true, but somehow there is a notion that using $\times$ and $\div$ to stand for multiplication and division is juvenile and childish and that other notations, like the ones I'm about to describe, are much more sophisticated and grown up.

No longer will you use the $\times$ for times. It's too easy to get it confused with a variable named $x$, and it's just not used all that much. If you're multiplying a number times a variable, for instance, multiplying 3 times $x$, you'll write them next to each other, like $3x$. Similarly, $5y$ means "five times $y$," and $xy$ means "$x$ times $y$." Because 35 means "thirty-five" and not "three times five," to get the multiplicative meaning you need to put a little dot centered vertically between the two numbers, like this: $3 \cdot 5$.

Another notation you'll see for multiplication will be putting things next to each other in parentheses. To multiply $2 \times 3$, you'll often see $(2)(3)$. The product of $x + y$ and $7x - 2y$ would be written $(x + y)(7x - 2y)$. In sort of a cross between this notation and the "just write things next to each other" one, you might see something like $8(15x - 2)$, which is just 8 multiplied by all of $15x - 2$. Don't worry that you haven't yet learned how to carry out multiplication of things with $x$ in them. The details of that are all spelled out later in this chapter.

Division will no longer be written with the $\div$ sign. It will almost exclusively be written as fractions or with a / sign. If you want to take $x$ and divide it by $y$, you'll either write $\frac{x}{y}$ or $x/y$.

The other operations you know are the same. Addition is still done with the plus sign; subtraction is still done with the minus sign. You use exponents and say that $x^3 = (x)(x)(x)$ or $x^3 = x \cdot x \cdot x$.

# Polynomials

There are several types of expressions you'll see again and again in math classes. The very simplest of these is called a monomial. A monomial is basically what you get if you take a variable, raise it to a whole-number exponent, and then multiply the whole thing by a number. Actually, you're allowed to use more than one variable raised to a power if you multiply them together. A monomial might look something like $9x^6$, $14x$, $x^8$, or $3x^2y^4$.

There is some jargon associated with monomials. The degree of the monomial is the number in the exponent. So $9x^6$ has degree 6 and $14x$ has degree 1. If there is more than one variable in the monomial, its degree is the sum of the exponents. So $3x^2y^4$ has degree 6. The number out in front is called a coefficient. With $14x$, the coefficient is 14; with $x^8$, the coefficient is 1.

It's okay to have the exponent on $x$ be 0. When you simplify something like $18x^0$, you'll get 18. While this might just look like a regular number to you, it is also a perfectly good monomial. The only rule about exponents in monomials is that they can't be negative and they must be whole numbers.

If you're up on your Greek prefixes, you might expect a polynomial to be made up of many monomials, and you'd be right. When you add together a bunch of monomials, what you get is a polynomial. A polynomial might look something like $7x^3 + 12x + 4$ or like $8 + x^5 + 19x^8$. There are a few things to keep in mind when looking at polynomials.

- Each of the monomials that make up the polynomial is called a term.

- Every exponent must be a whole number. No fractional exponents are allowed in a polynomial. No negative exponents are allowed in a polynomial.

- If there is a term with no visible $x$, it is called the constant term. Secretly there really is an $x^0$ in that term, but it is invisible.

- By tradition, we tend to write the terms of polynomials in order with the highest degree term first and working our way down to the constant term. Sometimes we start with the constant term and work our way up to the term of highest degree.

- The degree of a polynomial is defined to be the degree of its highest degree term.

- The coefficient of the highest degree term is called the leading coefficient.

**Example:** Is $x + x^{12} + 2x^0 + \pi$ a polynomial?

**Solution:** In order to decide if an expression is a polynomial, we need to make sure it is the sum of proper monomials. Let's go through this term-by-term and figure it out.

The first term, $x$, is secretly hiding both a coefficient and an exponent of 1. The whole-number exponent over our variable $x$ implies this is a fine term for a polynomial. We continue our inspection for rogue terms.

The second term, $x^{12}$ is also our variable raised to a whole-number power. So good so far.

The third term is a little more tricky, but the exponent 0 over $x$ is indeed a whole number. This is a valid monomial.

The fourth term has an invisible $x^0$ multiplied by it, implying that $\pi$ is a monomial, part of the constant term in fact.

Because each term of the four-term sum is a valid monomial, this is a valid polynomial.

**Example:** Is $\dfrac{1}{x} + 4x + 9x^2 + 16x^3$ a polynomial?

**Solution:** While each $x$ in this expression technically has a whole-number exponent at this point, it is not written as the sum of monomials. We need to try to rewrite this expression so that each term is of the form $c \cdot x^n$ for $n$ a whole number and $c$ a constant.

We're going to start by taking the first term out of fraction form. $\frac{1}{x}$ is equal to $x^{-1}$ by the rules for converting from fractional to exponential notation.

−1 isn't a whole number, so we know that the expression is not a polynomial.

> ### Practice Makes Perfect
>
> Problem 1: Determine if $x^2 + \sqrt[8]{2}x + \sqrt{x} + \pi$ is a polynomial.
>
> Problem 2: Determine if $x^5 + 5x^4 + 10x^3 + 10x^2 + 5x + 1$ is a polynomial.

# Combining Like Terms

Just like the phrase "oil and water don't mix," variables of different types don't want anything to do with each other when you're adding expressions together. Don't work against this natural tendency of variables and expressions. Instead, allow the different terms to express themselves. Those that go together shall be combined, and the ones that want to keep their distance should be allowed to stay separate.

You don't want to take this idea of separation too far, however. When there are things that are alike, they should be combined whenever possible to simplify the problem. Once you identify the like terms, simplifying them and combining them will allow you to take a monstrous piece of mathematical gibberish that extends across several lines and condense it into a tame expression with just a few terms.

You can add only $x$'s with other $x$'s, not $x^2$s or any other power of $x$. While $x$'s are related to $x^2$s, they're not necessarily equal. You can think of an $x$ as a frog egg, $x^2$s as tadpoles, and so on. They are all related to each other, but they're not the same. A frog egg plus a tadpole doesn't equal two tadpoles! However, 4 frog eggs + 5 tadpoles + 2 tadpoles + 1 frog egg = 5 frog eggs + 7 tadpoles. We can only take this analogy so far, but you probably now can see how $x + x + x^2 = 2x + x^2$. Because, like the eggs and tadpoles, you can't just add the $x$'s and $x^2$s together as if they were the same object.

So how do you go about combining terms? Let's imagine that you have five copies of $x$. It doesn't matter what $x$ stands for or represents in terms of the problem; all that matters is that you have five of them. You'd write this as $5x$. And what if you wanted to add this to seven more copies of $x$? Using math notation you would write this as $5x + 7x$. Can these be combined? Is there any way to simplify this?

You probably suspect that the answer to these questions is "yes," because why else would we ask (unless it was a part of a trick—but in this book there are no plans to trick you, as I have plenty of students of my own to trick). And you're right. Both of these terms only include one variable, plain $x$. Sure, there are coefficients on the $x$'s, but that doesn't change anything. The variable is $x$, and it's not acted on by exponents, roots, or other mathematical operations. So how do you go about combining a term with a plain $x$ with another term with a plain $x$?

Remember, when you multiply something by five, it's like having five copies of itself, so $5x$ is the same as $x + x + x + x + x$. And $7x$ would be $x + x + x + x + x + x + x$. When you add all these together, you'll get $x + x + x + x + x + x + x + x + x + x + x + x$, which, if you count up all the $x$'s, comes out to be $12x$, so $5x + 7x = 12x$. Probably you've noticed that $5 + 7 = 12$, and that you probably don't have to go through all the effort of writing out all the $x$'s, and you are absolutely right. If you are adding together two terms that have a coefficient multiplied by a plain $x$, you can combine them by adding together the coefficient.

**Example:** Add $18x + 7x$.

**Solution:** Add the coefficients. $18 + 7 = 25$, so $18x + 7x = 25x$.

This rule will work just as well if the coefficients are unpleasant numbers, if you are subtracting instead of adding (in which case you subtract the coefficients), and if your variable has a name other than $x$.

**Example:** Add $8.3x + 19.07x$.

**Solution:** Because $8.3 + 19.07 = 27.37$, $8.3x + 19.07x = 27.37x$.

**Example:** Subtract $4.2x - 6.1x$.

**Solution:** $4.2 - 6.1 = -1.9$, so $4.2x - 6.1x = -1.9x$.

**Example:** Add $3y + 2.4y$.

**Solution:** Because $3 + 2.4 = 5.4$, $3y + 2.4y = 5.4y$.

> **Practice Makes Perfect**
>
> Problem 3: Add $3.94x + 5.32x$.
> Problem 4: Subtract $234.2y - 563.3y$.

# Simplifying Products

You may be longing for the days when your multiplications were limited to just numbers. Now we've moved into the realm of having to multiply expressions by expressions. There's good news and bad news here. The good news is that for most of the problems you'll see of this type, the numbers are small enough that you shouldn't need to use a calculator. (And if you find yourself getting out your calculator to multiply $3 \times 8$, you really need to start brushing up on your multiplication tables!) The bad news is that you'll need to be careful to keep track of signs, order of operations, laws of algebra, and rules of exponents.

## Multiplying Monomials

The most basic way to combine expressions with multiplication is when you have one monomial that is multiplied by another. Remember that a monomial is an expression made up of a variable raised to a power and multiplied by a coefficient. Something like $-27x^{53}$ or the somewhat more pleasant-looking $3x^2$. Multiplying together two monomials is nothing more than applying the Commutative Law, the rules of multiplication, and the laws of exponents.

**Example:** Let's say you wanted to multiply together $8x^6$ and $-4x^3$.

**Solution:** Applying the Commutative Law to $(8x^6)(-4x^3)$, you can get the numbers together and the $x$'s together, like $(8)(-4)x^6x^3$.

You know that $(8)(-4) = -32$, giving you the coefficient for your answer.

> **Practice Makes Perfect**
>
> Problem 5: Multiply $4(7x^8)$.
> Problem 6: Multiply $(5x^3)(-9x^{20})$.

Remembering the rules of exponents, because you have the same base, you add the exponents, giving you $x^6x^4 = x^9$. Therefore, the answer will be $-32x^9$.

## Using the Distributive Law

Do you remember the Distributive Law? Are you extremely familiar with the Distributive Law and ready to use it with skill and aplomb in a variety of circumstances? If you answered "no" to either of those questions, go back to Chapter 3 and take a few minutes to review the Distributive Law. We'll wait. It's important. If there is a top-ten list of mistakes that students make, incorrect use of the Distributive Law definitely would make that list (although adding fractions probably would be at the top of the list).

The Distributive Law will allow us to simplify expressions like $7x(3x + 2)$ or anything where a single term is multiplied by things that are added together. Remember that the Distributive Law says that $a(b + c) = ab + ac$, and it doesn't matter whether $a$, $b$, and $c$ are numbers or variables or expressions made up of numbers and variables.

| Practice Makes Perfect |
| --- |
| Problem 7: Simplify $x^4(7x^5-x^2)$. |
| Problem 8: Simplify $(-5x)(3x^3-9x^5)$. |

**Example:** Use the Distributive Law to simplify $3x(7x + 12x^2)$.

**Solution:** By the Distributive Law, this is $(3x)(7x) + (3x)(12x^2)$.

Now you can use the rules of multiplying monomials to convert this to $21x^2 + 36x^3$.

## FOILed Again

A lot of things you'll see in math have stupid abbreviations to help you remember them. Some of them are so complicated that you're probably better off remembering the original rule than trying to figure out the mnemonic. In contrast, FOIL is short and to the point. It's the name given to the method for multiplying a *binomial* by a binomial. You'll use this all the time so it is worth knowing. There are some students who struggle with algebra who should give serious thought to having FOIL tattooed on their forearms. It's really that important.

A binomial is an expression of the form $a + b$ or $3x + 7y$ or $8x - 9x^2$. Anything where you have two monomials added together.

This procedure will work whether your binomials are made up of numbers and $x$ to different powers, they have two different variables, there are no variables but just pesky square roots, or there are other things that make multiplying awkward. Anything of the form $(a + b)(c + d)$ can be multiplied out using this method—no matter what $a$, $b$, $c$, or $d$ stand for.

**def•i•ni•tion**

A **binomial** is any sum of two monomials and it can include any combination of variables or constants.

The easiest way to learn the FOIL rule is to see it in action. Let's look at an example where we multiply together $(2x + 3)(4x^7 + x^2)$. FOIL stands for "First, Outer, Inner, Last" or else for "First, Outside, Inside, Last."

- First: You multiply together the first term from each set of parentheses. In this case, you'd multiply $(2x)(4x^7)$ and get $8x^8$. Write that down.

- Outer: Next you multiply together the terms on the outside, the first term from the first set of parentheses and the second term from the second set of parentheses. They're called the "outer" because they're the farthest away from each other and on the outer border of the problem. In this case you get $(2x)(x^2)$, which equals $2x^3$. You add this to the $8x^8$ you got from the previous part. Had there been a negative sign involved, you would have subtracted (more on this later).

- Inner: This is the product of the second term from the first set of parentheses and the first term from the second set. Here you multiply $(3)(4x^7)$, which is $12x^7$. Again, because everything in this problem is added, this is added, too.

- Last: Finally, it's the last times the last, so you multiply the second term from each binomial. Here it's $(3)(x^2)$, which is $3x^2$. This is added, too, because, say it with me, in this problem everything is added.

So your answer is $8x^8 + 2x^3 + 12x^7 + 3x^2$.

In order to multiply together two terms by two terms, you need to do four multiplications. There's really no way out of it except in some very, very special cases. Those are hardly worth mentioning now. The point being that FOIL is the way to do this type of problem, and you're stuck with it. Forgetting to use FOIL is hazardous to your grades.

So the first example was carefully concocted so that all the signs came out as plus signs. You won't be that lucky most of the time, as one or more of the signs is likely to be a minus sign. What do you do then?

The most straightforward way of handling a minus sign that happens between the two terms is to remember that subtracting is the same thing as adding the negative. You'll change the sign to "plus a minus" and then do the problem as you would before.

**Example:** Use FOIL to multiply $(3x + 1)(2x^2 - 8x)$.

**Solution:** To take care of the minus sign in the second factor, you could rewrite the problem as $(3x + 1)(2x^2 + -8x)$.

Now you can apply FOIL.

First: $(3x)(2x^2) = 6x^3$.

Outer: $(3x)(-8x) = -24x^2$.

Inner: $(1)(2x^2) = 2x^2$.

Last: $(1)(-8x) = -8x$.

Adding them all together: $6x^3 + -24x^2 + 2x^2 + -8x$.

To finish up the problem, there are two steps remaining. Convert the double signs back to regular signs. That is, no one wants to be left with an expression written as "plus a minus." It's polite to turn those back into regular subtraction. The other thing you'll need to do is to combine like terms.

So, finishing up, you'll have $6x^3 - 24x^2 + 2x^2 - 8x = 6x^3 - 22x^2 - 8x$.

Once you get practice at using FOIL when there are minus signs involved, you won't have to rewrite the problem with double signs. You'll be able to take care of it all in your head. But until you reach that point, you're better off writing everything down and doing all the problems the long way. It might seem like it's boring and a lot of work, but it's a heck of a lot less work than making a mistake and having to do the whole problem over again. Or, even worse, making a lot of mistakes day after day, week after week, and ending up being forced to take the entire course again!

The place where students forget to use FOIL is when the multiplication problem is written using the notation of exponents. If you see something of the form $(2x + 5)^2$, you need to use FOIL to expand it. The answer that you get from using FOIL is the only right answer to this problem. If you come up with a shortcut and it gives you a different answer, then it is wrong.

You need to use FOIL for problems like $(2x + 5)^2$ because this is really $(2x + 5)(2x + 5)$, which is to say two binomials multiplied together. Let's look at this problem and see how it's done.

**Example:** Expand $(2x + 5)^2$ with FOIL.

**Solution:** We start by writing this as $(2x + 5)(2x + 5)$, so it's clearer which terms we're talking about.

First: $(2x)(2x) = 4x^2$

Outer: $(2x)(5) = 10x$

Inner: $(5)(2x) = 10x$

Last: $(5)(5) = 25$

So your answer is $4x^2 + 10x + 10x + 25 = 4x^2 + 20x + 25$.

> **Practice Makes Perfect**
>
> Problem 9: Use FOIL to multiply $(7 + 3x)(5x^2 + 8x^3)$.
>
> Problem 10: Use FOIL to multiply $(8x - 5x^7)(11x^4 + 8x^3)$.

You might be wondering why FOIL works. There's nothing special about it; it's just an application of the Distributive Law.

# Plugging In: Numbers Return

In a lot of problems you'll want a number back at the end. After you've dealt with the mix of letters and numbers and have carefully conjured with them using the rules of algebra, you're going to come up with a number. One way that this happens is to assign numerical values to each letter in an expression and then to carry out the arithmetic. The fancy name for this is "evaluating the expression." You almost always will hear it called "plugging in."

It's not hard at all. You will be told what number each letter stands for, and then you replace every instance of that letter with its number. From then on, it's the same sort of stuff that you've been doing for a long time.

**Example:** Consider the expression $7x^2$. Evaluate this expression when $x = 6$.

**Solution:** "Consider the expression …" or "Consider the equation …" are a mathematician's way of saying, "Pay attention to this." You aren't expected to reach any deep meaning in life by thinking about it.

To evaluate the expression, every place you see an $x$, you replace it with a 6. You'll get $7(6)^2 = (7)(36) = 252$.

**Example:** In the expression $\dfrac{14y - 2x^2}{15x}$, evaluate it with $x = 2$ and $y = 4$.

Problem 11: Evaluate $2x^3$ when $x = 5$.

Problem 12: Evaluate $(-5x)(3x^3 - 9x^5)$ when $x = 4$.

**Solution:** The most complicated part of this problem is keeping track of the parentheses and the order of operations. Plugging in, you get $((14)(4) - 2(2)^2)/((15)(2))$.

Using order of operations and doing the multiplications inside the parentheses, you'll get $(56 - 8)/30 = 48/30 = 1.6$.

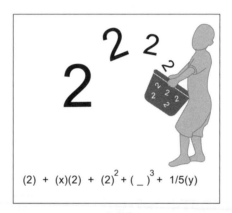

$(2) + (x)(2) + (2)^2 + (\_)^3 + 1/5(y)$

# Found in Translation

In pre-algebra, the whole point of bringing letters like $x$ and $y$ and such into the mix and turning problems with numbers that you could check on a calculator into arcane chunks of symbols that you can't really do anything with yet is because it turns out to be really useful for solving a lot of problems. Algebra, which is really what you're heading into here, has been around for over a thousand years because it can be used to solve so many problems. Changing English into math is usually more complicated than changing math into English, so we'll start by unraveling the mysterious meaning of some of these equations.

When you're taking a mathematical expression and turning it back into English, it's less of a precise art. There are a lot of ways to say the same thing in English, and there isn't necessarily one right way. Also, the meaning will depend on whether the mathematical situation you're looking at is abstract or tied to a real-world example.

**Example:** Write a statement in English that represents: $4x$.

**Solution:** Four times as much as a number.

**Example:** Write a statement in English that represents: $\frac{1}{5}x - 5$.

**Solution:** Five less than one-fifth of a number.

While you might doubt the usefulness of some of the examples you might see in math and science books (such as the two previous ones in this book), anyone from the United States who travels to anywhere else in the world can make use of this one. People who travel to the United States will have to do a related problem. While temperature in the United States is measured in Fahrenheit, the rest of the world uses Celsius. An American arriving in another country in the middle of summer might be told that it is 28 degrees and will probably want to figure out how warm that really is and why nobody is wearing a coat.

The formula for converting temperature from Celsius to Fahrenheit says that $F = \frac{9}{5}C + 32$. Given a Celsius temperature, you plug it in for $C$ and evaluate the expression to get $F$.

**Example:** Translate the temperature conversion formula into English.

**Solution:** The temperature in Fahrenheit is equal to nine-fifths the temperature in Celsius plus 32.

Now let's use the temperature conversion formula to convert the current temperature of 28 degrees Celsius in Hong Kong on September 6th, 2007, at 7:50 P.M. local time into Fahrenheit.

**Example:** It is 28 degrees Celsius. What's the temperature in Fahrenheit?

**Solution:** In the expression $\frac{9}{5}C + 32$, you plug in 28 for $C$. You'll get $\frac{9}{5}(28) + 32 = 82.4$. A nice evening, but it might be a little bit warm.

> **Practice Makes Perfect**
>
> Problem 13: Write a statement in English that represents: $11 + 11x$.
>
> Problem 14: Write a statement in English that represents: $(-5x)(3x^3)$.

> **Practice Makes Perfect**
>
> Problem 15: The formula for converting temperature from Celsius to Fahrenheit says that $F = \frac{9}{5}C + 32$. Given a Celsius temperature, you plug it in for $C$ and evaluate the expression to get $F$. Translate the temperature conversion formula into English.

## The Least You Need to Know

◆ Variables are letters that stand in for numbers.

◆ A mathematical expression is any quantity made up of numbers, operations, and variables.

◆ FOIL stands for "first, outer, inner, last" and is used for multiplying binomials.

◆ Many problems can be solved by replacing the variables with specified numbers and evaluating the expression according to the normal rules of arithmetic.

# Introduction to Equations

## In This Chapter

- ◆ Solving basic equations

- ◆ Equations with many variables

- ◆ Checking your work

Algebra was invented about 1,200 years ago in the part of the world that is now called the Middle East. An Islamic scholar by the name of Al-Khwarizmi wrote a book whose title roughly translates to "The Compendious Book on Calculation by Completion and Balancing," and the technique he called "completion" was known by the name al-jabr, from which the modern word for algebra derives.

People will debate what exactly algebra is, but one of the core techniques in the subject is finding the solutions to equations by applying the laws of math. Think of it as a puzzle made up of an expression with one variable in it as well as some other information. Your task is to figure out what number the variable is equal to.

There's much more to algebra than finding the solutions to equations, but that's a matter for another course, namely algebra. And if you can't get enough algebra in this book, you'll have plenty of opportunities to study it further.

# What Are Equations?

Up to this point you've worked with numbers and expressions. In the grammar of mathematics, these have been fairly simple utterances without verbs, like the way you might hear cavemen on a cartoon talk. Three plus seven! Four times *x* minus six! See, no verbs. Most of the problems have come with written instructions that gave you more information.

**Amy's Answers**

An equation is a sentence with the verb equals in it. Whatever is on one side equals whatever is on the other. To keep this sentence true (you wouldn't lie to a math teacher, would you?) you need to do exactly the same thing to both sides when you're changing the equation.

**Kositsky's Cautions**

It is really, incredibly, unbelievably important that you never abuse the equals sign. Only use the equals sign when you know both sides are equal to each other. If you subtract, add, multiply, or divide one of the sides by something, you need to start on a new line with a new equals sign not connected to your old equals sign!

Equations are mathematical sentences where the verb is "equals." They talk about two quantities and assert that they are equal. The equals sign = is used to show this relationship. The equals sign is not to be used lightly. It can be used only when two quantities have exactly the same value and can be used interchangeably. Not when things are only related to each other. Not to show that you've moved on to the next stage of a problem. You can use an equal sign only when things are precisely and totally equal. This may seem obvious now, but it's a mistake students tend to make.

An equation is any mathematical statement that has two equal quantities joined by an equals sign. It can be something fairly simple like $2 + 3 = 5$ or rather complicated like $y^2 = x^4 - x + 1$. In either of these cases, the expression on the left-hand side of the equals sign has exactly the same value as the expression on the right-hand side of the equals sign. In the first case, it's really obvious. In the second case, you'll have to take our word for it right now.

The equations that you'll be working with in this chapter will be relatively straightforward. Most will look something like $x - 3 = 9$ or $6x = 12$ or maybe $6x + 9 = 27$. They will only have one variable in them (and it will typically be named *x*, but it doesn't really matter what its name is).

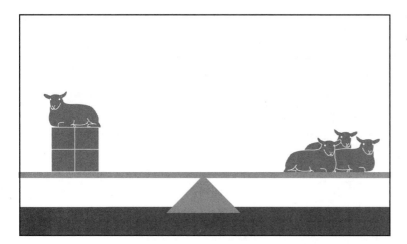

*The see-saw balances when both sides have equal values.*

Your task will be simple: find the one single value of *x* that makes the equation a true statement. For each equation of this sort, there is exactly one real number that you can plug in for *x* that makes it a true statement. Any other value would make it a terrible lie. If you look at an equation like *x* – 2 = 4, you probably can see that if you plug in 6 for *x* you will get 6 – 2 = 4, which is a true statement. Therefore, 6 is the value of *x* that makes this equation true. This is called a *solution to the equation.* Any number that you can plug in for the variable to make the equation true is a solution to the equation. This process of finding the solution is called solving the equation. If you pick any number other than 6, like 3, and plug it in, you'll get something false. Plugging in 3 for *x* gives you 3 – 2 = 4, which is clearly nonsense and mathematical crazy-talk. You don't even want to go there.

You'll be responsible for the honorable task of finding that one value and preserving the integrity of mathematics by making sure that the equations don't lie. There are two main ideas that you will use when solving equations.

The first idea hinges on the fact that both sides of the equation are exactly and precisely equal. If you make a change to one side of an equation, you have to make the same change to the other side of the equation. If you add 1 to the left side, you'll have

## def•i•ni•tion

The **solution to an equation** is a collection of values for x, y, and other variables that make the equation true.

For example, 2x = 3 has the solution x = 1.5 because 2(1.5) = 3. And 3 is not a solution to 2x = 3 because 2(3) = 3 is false! 6 and 3 can't be equal. We write this mathematically as 2(3) ≠ 3.

to add 1 to the right side. If you multiply the right side by 3, you'll have to multiply the left side by 3. No matter what you do to one side of the equation, you must do the exact same thing to the other side. Imagine that an equation is like a see-saw that's perfectly balanced. If a kid gets off one side, then that side is going to fly up unless a kid gets off the other side.

The second idea you will use is inverse operations. This is the method that will help you decide what to do to the sides of the equation. Doing things willy-nilly to both sides of the equation, while mathematically correct, will usually get you no closer to the correct solution of the equation. You need a strategy and the inverse operations are what you're going to use.

Just like all the other properties and laws that have come back to haunt you in algebra, inverse operations are no exception. It is painfully obvious: if you add something to a number and then subtract it, you end up right back where you started. However, this simple operation is one of the key ideas that will appear in algebra problems. As will the related idea that if you multiply by a number then divide by the same number, you'll return to where you started. If you have $x - 3 = 7$, you need to add 3 to both sides to undo the subtraction of 3 on the left-hand side of the equation. It's the same thing if you raise a number to a power and then take that root. Just remember that inverse operations are like "undo" and you'll make the right choices.

# Equations with One Type of Operation

Just like you have to crawl before you can walk, you'll have to learn how to solve equations with one operation before you can move on to more complicated equations. We'll start with a simple equation and ask you to solve it: $x - 1 = 3$. Probably without resorting to inverse operations, to doing the same thing to both sides of the equation, or to anything else that has been described over the past few pages, you have come up with the answer $x = 4$.

**Kositsky's Cautions** _____

When solving the equation $x - 1 = 3$, the answer was given as $x = 4$. This is the best way to write the answer, rather than just plain 4. You want to make it clear what you mean about 4. 4 what? What about 4? Writing out $x = 4$ is the mathematical equivalent of giving your answer as a complete sentence.

If you were asked how you did it, you would probably say that it was obvious. That you just knew it. That if you took a number and subtracted 1 from it and got 3 as your answer, then the only possible value for the number has to be 4. These may all be true, but please, please, please follow the steps described in the next few pages for solving equations. When equations get more complicated, you'll want to be a master at applying these methods and if you skimp on the practice now, you'll be at a disadvantage.

And what if the number 4 didn't just jump magically into your head? What if you're thinking, "Oh, no, they expected us to know that it was 4 and I have no idea. This is bad." Don't fear. It's actually good that you don't have any pre-conceived notions of how these problems should be done. You don't have to unlearn any bad habits. You can learn how to do things the right way from the start.

The goal of solving this type of equation is to get $x$ by itself on one side of the equation, typically the left. It's fine to go against tradition and try to get your $x$'s on the right, but just know that you're being a bit countercultural. And as you work through a problem, your work might not look like other people's. But it's fine. You don't need to stay with the pack. You don't need to be like everyone else. You can be your own person mathematically and put your variables on the right side of the equals sign.

The method you'll use was foreshadowed before. Did you hear the scary music when inverse operations were mentioned? This is why. You'll be using inverse operations to strip all the numbers away from your variable and to leave it alone and isolated on the left side (or, if you swing that way, the right). Just remember that whatever you do to one side of the equation has to be done to the other side of the equation (no getting around this rule).

## Addition and Subtraction

Just like you learned how to add and subtract before you learned how to multiply and divide, you'll start with equations where the only operations are addition and subtraction. And you remember that addition and subtraction are secretly sort of the same thing because subtracting is the same as adding the negative, right?

Let's start by solving $x - 8 = 2$. This is clearly an equation because it has an equals sign between some mathematical stuff. There is only one operation, minus, so this equation seems to fit into our category of solving equations with one operation where the operation is addition or subtraction. So you're in the right place. Now what?

Here's where the inverse operations come in. How do we undo minus? With plus! If you don't remember this, see Chapter 3. You're trying to get that – 8 to go away and leave $x$ alone. So what to do? The way that you undo subtracting 8 is to add 8. And since you must do the same thing to both sides, you will add 8 to both sides.

This gives you something like $x - 8 + 8 = 2 + 8$.

Simplifying, you're left with $x = 10$. And that's your answer.

**Example:** Solve the equation $x + 6 = 2$.

**Solution:** Again, this is an equation with only one operation, which is plus or minus. Here the 6 is added, so the inverse operation is to subtract it from both sides.

This gives you $x + 6 - 6 = 2 - 6$.

Simplifying, you're left with $x = -4$, which is your answer.

> **Practice Makes Perfect**
>
> Problem 1: Solve the equation $x - 7 = 4$ for $x$.
>
> Problem 2: Solve the equation $x + 2 = 17 - 6$ for $x$.

## Multiplication and Division

What if instead of taking $x$ and adding something to it or subtracting something from it, we multiply it by something. Can we retrieve the original $x$ and get it to tell us its value, the one single number that is the solution to the equation? Of course we can, because we know that multiplication and division are inverse operations and that we can undo multiplication by dividing.

**Example:** Solve the equation $2x = 14$.

**Solution:** At this point you might be thinking what the answer is and starting to wonder if there is any reason to use the long way. But there is; bear with us.

Because $x$ was multiplied by 2, in order to isolate the variable we need to get that 2 away from it. Remember, the 2 and the $x$ are not connected by crazy glue or by undying love, but rather just by multiplication. What has been joined together by multiplication can be undone by division. To get $x$ un-multiplied by 2, you will divide both sides of the equation by 2.

This gives you $\frac{2x}{2} = \frac{14}{2}$.

Simplifying, this is $x = 7$.

The only catch to this is if your $x$ is multiplied by a fraction. It's not even really a catch, it's just something that you need to be a little bit careful with. You'll need to divide by the fraction. Just think back to the arithmetic of fractions and remember that to divide by a fraction, you multiply by its reciprocal.

**Example:** Solve $\frac{3}{7}x = 6$.

**Solution:** To undo multiplying by $\frac{3}{7}$, you multiply both sides of the equation by $\frac{7}{3}$, giving you $\frac{7}{3} \times \frac{3}{7}x = \frac{7}{3} \times 6$. Now when you simplify, you're left with $x = 14$.

> **Practice Makes Perfect**
>
> Problem 3: Solve $\frac{5}{29}z = 5$ for $z$.
>
> Problem 4: Solve $\frac{3}{7} = 6x$ for $x$.

# Multi-Step Equations

As you know, life is not always simple and neither are equations. They will continue to get more complicated as you have to take more steps in order to solve them. If you keep your wits about you, though, even the tricky problems will fall at your hands.

## More Than One Operation

Sometimes $x$ might be entangled with a few different numbers. It might be multiplied by something and then have something else added to it. What to do in this situation? Just approach it step by step and undo each of the operations, remembering that whatever you do to one side of the equation needs to be done to the other side of the equation.

As a rule of thumb, start by undoing any additions or subtractions and then deal with the multiplications and divisions. Why is this the way it is done? I could make a vague reference to the order of operations. I could make some sort of clever mathematical argument. When it comes down to it, the real answer is because it is easier. If you do it in this order, you will not have to deal with complicated rules for dealing with expressions with fractions and you'll be less likely to make a mistake. Ignore this advice at your own risk.

**Example:** Solve $2x - 5 = 11$.

**Solution:** Remembering the warning from above, you will start by undoing the subtraction. Adding 5 to both sides gives you $2x - 5 + 5 = 11 + 5$ or $2x = 16$. Now you're in a familiar situation. You can now divide both sides by 2, leaving you with $x = 8$.

> **Practice Makes Perfect**
>
> Problem 5: Solve $\frac{5}{29} + 4a = 5$ for $a$.
>
> Problem 6: Solve for $x$.
> 1) $5 = 1 - \frac{x}{2}$
> 2) $2x + 7 + 4x = 18$
> 3) $5 - 3x = 2$
> 4) $3x - 9 = 9$

# Combining Like Terms

What if the letters and numbers are all jumbled on both sides of the equation? You'll have to do something to get them on the same side.

When you're asked to solve something like $2x + 4 = x + 7$, will the previous techniques still work? Yes. You'll start by combining like terms by carrying out all the additions and subtractions on each side of the equation. After that, you'll move the $x$'s to one side and the numbers to the other, following the methods that you've already practiced.

**Example:** Solve $2x + 4 = x + 7$.

**Solution:** In this problem, you want to get the $x$'s on one side (I'm picking the left) and the numbers on the other side (the right). Working with numbers is a little bit less intimidating than working with letters, so start by trying to eliminate the 4 from the left-hand side (the side that is supposed to have only letters).

Now, in this case 4 is added on to the $2x$, so the only way to make the + 4 go away is to subtract 4 from both sides, giving you $2x + 4 - 4 = x + 7 - 4$.

*If you subtract the same thing from both sides, the see-saw still balances.*

Simplifying, you'll have $2x = x + 3$. This brings you closer to the goal, but you're not quite there. You need to get the $x$ off the right side of the equation. What you will do is subtract $x$ from both sides. Up to this point you have been adding and subtracting numbers on both sides of the equation. You can do this with $x$'s, too. It works the same way.

Subtracting gives you $2x - x = x - x + 3$. Combining like terms on the left and right, you'll have $x = 3$, which is the answer.

**Example:** Solve $3x + 2 + x = 6 + 2x + 8$.

**Solution:** Start by combining like terms on each side of the equation. Here you would get $4x + 2 = 2x + 14$.

| Practice Makes Perfect |
| --- |
| Problem 7: Solve $a + 4a = 17$ for $a$. |
| Problem 8: Solve $3 - \frac{z}{6} = 8 + z$ for $z$. |

Now move the numbers to one side and the letters to the other side by using inverse operations and remembering to always do the same thing to both sides of the equation. Start by subtracting 2 from both sides, giving $4x + 2 - 2 = 2x + 14 - 2$, which is $4x = 2x + 12$. Now you subtract $2x$ from each side, for $4x - 2x = 2x + 12 - 2x$. Again, combine like terms to get $2x = 12$. This is solved as $x = 6$.

# Equations with More Than One Variable

As you might expect, the $x$'s don't mix with the $y$'s, and the $y$'s don't mix with the $z$'s, and every letter in an expression sticks with its own kind. Yet, you can combine $x$'s with other $x$'s and $y$'s with other $y$'s—just as long as you follow a few simple rules.

What happens if a problem has both $x$'s and $y$'s in it? What if the writer of the problem strays from the conventions of choosing variable names and fills your problem with an entire alphabet soup of variables to be added and combined? Don't worry: you just deal with each variable separately. You'll combine terms that have the same variable until there is only one term with each variable left.

**Example:** Solve $0 = 5x - 3y - 4$ for $x$.

**Solution:** This problem has two variables in it: $x$ and $y$. Because we are asked to find $x$, we want to move all the $x$'s to one side of the equation and everything else to the other. It doesn't matter which side you want to put the $x$'s on, but we'll choose the right in this example.

All of the $x$'s are already on the right, but we need to get rid of the pesky $y$'s and any constant terms. First, let's undo the subtraction of 4 by adding 4 to both sides of the equation.

$$0 + 4 = 5x - 3y - 4 + 4$$

$$4 = 5x - 3y$$

Now we need to undo the subtraction of $3y$ by adding $3y$ to both sides of the equation.

$$4 + 3y = 5x - 3y + 3y$$

$$4 + 3y = 5x$$

Finally, we need to divide by 5 to get the coefficient of $5x$ to be 1.

$$\frac{4+3y}{5} = \frac{5x}{5}$$

$$\frac{4+3y}{5} = x$$

$x$ is equal to an expression with a variable in it, but that's sometimes as good as it gets!

**Example:** Simplify $7x + 14y - 3x - 2x + 3z + 6y$.

**Solution:** This problem has three variables in it: $x$, $y$, and $z$. To begin with, we need to keep them separated. A good first step is to write the problem so that the terms for each variable are grouped together next to each other. In order to do this, you have to be very careful to observe the rules of algebra. If all that you were doing was adding, you could rearrange things without a second thought by using the Commutative Law. Did you think that those laws were taught just to torment you? No, they come up all the time when working with expressions and simplifying them. Remember that subtraction is the same as adding a negative and you'll be alright.

Because this example doesn't have any parentheses in it, you can use the Commutative Law to rearrange the terms so that the variables are grouped together. You can put the variables in any order you want. Alphabetical order is a good choice. If they turn out to spell a word, you can choose that as your order—unless the word is impolite and the person who'll be reading your work is uptight. In that case, you'll probably want to stick to some other order.

Rearranging the terms, we get $7x - 3x - 2x + 14y + 6y + 3z$.

Using the order of operations, we work from left to right,

$$= (7x - 3x - 2x) + (14y + 6y) + 3z$$

$$= (4x - 2x) + (14y + 6y) + 3z$$

$$= 2x + (14y + 6y) + 3z$$

$$= 2x + 20y + 3z$$

At this point there is nothing further to be done. You can't combine things any more.

**Example:** Simplify: $8c + 4b - 34r + 19c - 10r + 50r - b + c - 3f + 8q + f$.

**Solution:** Start by rearranging the terms so that each letter is together:
$4b - b + 8c + 19c + c - 3f + f + 8q - 34r - 10r + 50r$.

$= 3b + 8c + 19c + c - 3f + f + 8q - 34r - 10r + 50r$

$= 3b + 27c + c - 3f + f + 8q - 34r - 10r + 50r$

$= 3b + 28c - 3f + f + 8q - 34r - 10r + 50r$

$= 3b + 28c - 2f + 8q - 34r - 10r + 50r$

$= 3b + 28c - 2f + 8q - 44r + 50r$

$= 3b + 28c - 2f + 8q + 6r$

> **Practice Makes Perfect**
>
> Problem 9: Solve $5a - {}^b/_4 = 10x + z$ for $x$.
>
> Problem 10: Solve $-3z - {}^{5a}/_{15} = 8a + yx^3$ for $a$.

# Equations with More Than One Solution

So far in your mathematics career, there has probably been only one correct answer to each problem. Those days have passed, and from here on out mathematics gets more complicated. Here's a classic.

**Example:** Find $x$ such that $x^2 = 1$.

**Solution:** What number squared is 1? A keen intuition or some sharp guesswork will probably tell you that $x = 1$ because $1^2 = 1$. Others might decide that $x = -1$ because $(-1)^2 = 1$. So which is the correct answer? Both of these answers by themselves are right and wrong! This is possible because I am a mean mathematician attempting to drive my poor victims … er … readers crazy. This is also true and false.

The example question is intentionally vague. The question can be read as, "Find [one] $x$ such that $x^2 = 1$." If this is the case, either "$x = -1$" or "$x = 1$" are correct. However you also can think of the question as "Find [all] $x$ such that $x^2 = 1$." If your teacher wants the second interpretation, neither "$x = -1$" nor "$x = 1$" is the correct answer. Your teacher probably hopes you will write $x = 1, -1$. This means that $x$ can equal either 1 *or* –1. You don't have enough information to say anything more about what $x$ is, so the proper solution is to list the two possible values for $x$. If this comes as a major shock to you, feel free to sit back and recover for a minute while the idea rolls around in your mind. This is an equation that has two solutions.

Isn't math supposed to be the class where there's only one right answer? English, art, social studies, P.E., even sometimes science regularly let you have the freedom to be creative, to choose between alternatives. You never had to turn in the One Right Essay (though it may have felt like it). If you think about it, your teacher probably would have been angry rather than proud if everyone in the class turned in the exact same essay, no matter how perfect that essay is! Well, however sad it may be, it's true that some problems in mathematics have more than one solution. In fact, you can get an infinite number of solutions to some problems and no solutions to others! You'll get to see these oddities in your algebra course when you continue on in mathematics.

## Polynomials Reloaded

Often, your problems are going to contain just one variable, almost always $x$, just raised to different powers. You'll find yourself working with polynomials and having to combine like terms to simplify them.

Here you need to keep in mind that while plain $x$ terms can be added to other plain $x$ terms, you can't simplify or combine terms that have different powers of $x$. If the terms have the same power of $x$, however, you can add or subtract them by adding or subtracting the coefficients.

**Example:** Add $6x^2 + 12x^2$.

**Solution:** Adding the coefficients, much like you did before, you get $18x^2$.

**Example:** What if you were asked to simplify $6x - 3x^2$?

**Solution:** There's nothing that you can do to simplify that expression. The different terms have different powers of $x$, so they can't be combined.

**Example:** Simplify: $7x^3 - 2x^9 + 3x^7 + 7x^3 - 5x^6 + 3x^6 + 6x^5 + 4x^9$.

**Solution:** Probably the best way to start is by putting the terms in order based on the exponents of $x$.

This gives you: $-2x^9 + 4x^9 + 3x^7 - 5x^6 + 3x^6 + 6x^5 + 7x^3 + 7x^3$.

Combining like terms,

$$= 2x^9 + 3x^7 - 5x^6 + 3x^6 + 6x^5 + 7x^3 + 7x^3$$

$$= 2x^9 + 3x^7 - 2x^6 + 6x^5 + 7x^3 + 7x^3$$

$$= 2x^9 + 3x^7 - 2x^6 + 6x^5 + 14x^3$$

---

**Practice Makes Perfect**

Problem 11: Simplify $1x^2 - 2x^2 + 3x^2 - 4x^2 + 5x^2 - 6x^2 + 7x^2 - 8x^2$.

Problem 12: Simplify $32x^7 - 12x^1 + 15x^4 + \frac{1}{2}x^1 - 3x^1 + 5x^2 + \frac{3}{4}x^5 + 4x^6$.

# Multiple Variables with Various Powers

We can play the same game of combining terms with multiple variables. The rules are the same, but you can't combine terms that don't have all the same variables with all the same powers in the term. This is somewhat confusing because the power on the variable $x$ doesn't have to be the same as the power on the variable $z$. The $x$'s exponent only needs to equal the other term's exponent of exponent. Let's jump into examples so you can get a feeling for what this jumble of words means.

**Example:** Add $6yxz^2 + 12xyz^2$.

**Solution:** Start by separating the terms completely so you know how many you have and what they are. In this case, the first term is $6yxz^2$ and the second term is $12xyz^2$.

Look at the variables in each term. Both the first and second term have $x$'s, $y$'s, and $z$'s, though they're in different orders.

Let's rearrange one of the terms so the variables are in the same order in both. By the Commutative Law of multiplication, we're going to replace the product $yx$ with the product $xy$. You can think about these rearrangements as changing the order of purchasing items at a supermarket. Whether you put the bananas or the pears first on that sliding black thingy, you still have to pay the cashier to avoid theft. The first term becomes $6xyz^2$.

Now we need to compare the exponents on each of the variables. The $x$'s of both terms have an implicit exponent of 1 because they have no number to the upper right of them. The same is true of the $y$'s; they, too, have implicit exponents of 1. The $z$'s both have exponents of 2.

That's all the variables. Their exponents all match correctly, so we can add these two terms by adding their coefficients and placing the variables next to the sum. This gives $6xyz^2 + 12xyz^2 = 18xyz^2$ with the right-hand side of the equation, $18 \, xyz^2$, as our answer.

**Example:** Add $6x^2 + 12yx^2$.

**Solution:** You may see the $x^2$s and be tempted to add the coefficients, but don't be too hasty. First we need to make sure all variables match up properly.

$x$ is in both terms and it is raised to the second power in both terms. So far so good.

We need to compare the power of the variable $y$ in the two terms, but $y$ doesn't appear in the first term! Here is one of those times where the magical number 1 comes to the rescue. Because anything raised to the zeroth power is equal to 1, $y^0 = 1$.

We can rewrite the problem after multiplying $6x^2$ by our cleverly written 1, $y^0$, and writing the implicit exponent of 1 as $6y^0x^2 + 12y^1x^2$.

It's now clear that the exponent over the variable $y$ is different in each term. Thus, no terms can be combined in this expression.

**Example:** Add $6x^2 + 12yx^2 + 12x^2y + 5x^3$.

**Solution:** First we need to look at each term and find out which ones have the same variables. Let's start with the first term. $6x^2$ has only one variable, $x$, and $x$ is raised to the second power. The second and third terms each have two variables, $x$ and $y$. If the variables themselves are different, you can't add the terms together because one of them is completely missing. Therefore, we cannot combine the first term with the second or the third.

> **Timely Tips**
>
> If one term has a variable with a non-zero exponent and a second term is missing that variable or that variable has a zero exponent, you cannot combine terms.

The remaining term, $5x^3$, has an exponent of 3. We can't combine two terms that have different exponents on the same variable, so there is nothing we can combine the last term with. The second term has two variables, $y$ and $x$, with $y$ raised to the first power and $x$ raised to the second power. We now want to see if we can add it to the third term, $12x^2y$, which has $y$ raised to the first power and $x$ raised to the second power. The exponents are the same! This means we can combine these two terms by adding their coefficients: $12yx^2 + 12x^2y = 24x^2y$.

We already know the fourth term cannot be combined with the first term. The fourth term doesn't have a $y$ in it, so we can't combine it with the second-plus-third term.

Our final answer with all terms combined is $6x^2 + 24yx^2 + 5x^3$.

**Example:** Simplify: $7xzy - 2zxy + 3x^3 + 7yzx - 5zx^2y + 3x^6 + 6x^5 + 4x^9$.

**Solution:** While this problem may look more complicated than the one before it, it's not that bad once you break it down into steps. A great place to start is by alphabetizing the variables that appear in each term. This isn't some sort of OCD behavior; it actually helps you solve the problem.

You should then have $7xyz - 2xyz + 3x^3 + 7xyz - 5x^2yz + 3x^6 + 6x^5 + 4x^9$.

Now you can start combining like terms. The terms that all have an $xyz$ can be combined. Doing it step by step, you'll get $5xyz + 3x^3 + 7xyz - 5x^2yz + 3x^6 + 6x^5 + 4x^9$ when you combine the first two terms.

Next you'll combine the other terms with $xyz$, resulting in $12xyz + 3x^3 - 5x^2yz + 3x^6 + 6x^5 + 4x^9$.

Is there anything else you can do to simplify? What about all those terms that have $x$ in them? Nope! Don't even try it! The terms with $x$ all have $x$ to different powers, so they can't be combined. You're done.

<table>
<tr><td colspan="1">Practice Makes Perfect</td></tr>
<tr><td>Problem 13: Simplify $17xy^0 - 2xz^1 + xz^3 - 3xy^1 + 5xa^0 + 10xz^4 + zx^6 + \frac{5}{2}xz^1$.<br><br>Problem 14: Simplify $5^0 + 10^5 + z^6 + 17xy^0 - 2z^6 - 2xz^1 + xz^0 + xy^4 - 3x^1 + 13xz$.</td></tr>
</table>

## Beyond Polynomials

If you are enrolled in a course designed by evil math geniuses who want to shove every little bit of math knowledge into your mind, you may have to learn how to simplify all manner of complicated expressions by combining like terms. You might be faced with square roots, fractional powers, negative powers, and other arcane mathematical invocations. Never fear. They're just trying to psych you out, and you can use the same techniques that worked for polynomials. Just take your expression, write every term as a coefficient times $x$ (or whatever your variable is) raised to the appropriate power, and simplify. This may require you to rewrite roots as fractional exponents and reciprocals as negative exponents, but you already know how to do that part. You'll see how to deal with all these complications and how they fit in with equations in the next chapter.

**Example:** Simplify $\dfrac{x^4}{\sqrt[5]{y}} \cdot x \cdot y^{\frac{7}{9}} = y \cdot 2x^{-\frac{3}{4}}$, putting all the $x$'s on one side of the equation and all the $y$'s on the other.

**Solution:** Wow, that seems like a doozy. Never fear, this problem can be tackled if you work step by step. The key to the entire problem is to remember that it hinges upon the laws of exponents.

Start by taking the radical in the denominator and writing it in terms of exponents. Since it's a fifth root, you'll be looking at the $\frac{1}{5}$ power; since it's in the denominator, it's a negative power.

This lets us write the equation as $x^4 \cdot y^{-\frac{1}{5}} \cdot x \cdot y^{\frac{7}{9}} = y \cdot 2x^{-\frac{3}{4}}$.

Next you want to group all the $x$'s and all the $y$'s. This converts the equation to $x^4 \cdot x \cdot y^{\frac{7}{9}} \cdot y^{-\frac{1}{5}} = y \cdot 2x^{-\frac{3}{4}}$.

Next, apply the rules of exponents to each side. Remember, when you have like bases, you add the exponents. You get $x^{4+1}y^{\frac{7}{9}-\frac{1}{5}} = y \cdot 2x^{-\frac{3}{4}}$. Simplifying (remember, you need to get a common denominator to subtract the fractions), this is $x^5 \cdot y^{\frac{26}{45}} = y \cdot 2x^{-\frac{3}{4}}$.

Almost done! To get the $x$'s on the left side of the equation, you need to multiply both sides of the equation by $x^{3/4}$. To get the $y$'s on the right side, you need to multiply both sides by $y^{-26/45}$.

Now you have $x^5 \cdot x^{\frac{3}{4}} \cdot y^{\frac{26}{45}} \cdot y^{-\frac{26}{45}} = y \cdot y^{-\frac{26}{45}} \cdot 2x^{-\frac{3}{4}} \cdot x^{\frac{3}{4}}$. Whoa, is this problem looking worse before it gets better? Don't worry.

Again, you add the exponents when you have like bases. Fortunately, now things will start to cancel.

Now you should have $x^{5+\frac{3}{4}} \cdot y^0 = y^{1-\frac{26}{45}} \cdot 2x^0$. That looks a lot better, no?

Now remember that anything to the zeroth power is 1, and everything simplifies to $x^{\frac{23}{4}} = y^{\frac{19}{45}}$.

# Checking Your Work

Allow me to tell you about a very special opportunity that you have when working with equations. You have the chance to make sure that you really do have the right answer. When you were working with plain numbers and with expressions, the only way that you had to make sure that you did a problem right was to do it again. But if you made a mistake the first time (or weren't quite sure how to do it), you would be likely to make the same mistake the second time through. Not much of a way to catch errors.

With equations, there is a way for you to check your work. With the equations that we are dealing with, there is only one right answer. (As things get more complicated, there will be more right answers. Life is like that.) And if you plug that answer back in for $x$ in the original equation, you should get a true equation as a result.

## Single Variable

Remember how you solved $2x - 5 = 11$ and you got $x = 8$ as an answer? You don't have to trust your skills at algebra that you got the right answer on the first try. Although you might suspect that that was the case, as you were being guided along. You can plug that $x$ value back into the equation. If you get a true mathematical statement, then you know you did the problem right. If you get something untrue, then you must have made a mistake somewhere and you should double-check both your algebra steps as well as how you plugged in.

**Example:** Check that the answer $x = 8$ is a solution to $2x - 5 = 11$.

**Solution:** You plug in 8 everywhere that there is an $x$ and get the statement $2(8) - 5 = 11$.

This simplifies to $16 - 5 = 11$, which is true because $11 = 11$. Since you get a true statement, you know that your answer is right.

It might seem tedious and like extra work to check your answers, but it's a good practice (and required in some classes). If you catch your mistakes before your teacher does, you can fix them and avoid having points taken off.

> **Practice Makes Perfect**
>
> Problem 15: Check that $x = -2$ is a solution to $2 + 12x = x^5 + x^4 + x^3 + x^2 + x^1$.

## Multiple Variables

These situations are a bit trickier, but it's still possible to check your work when you have multiple variables. You just need to plug in the expressions that the variables are equal to into the original equation.

**Example:** Earlier, we solved the equation for $x$ and arrived at the answer $\dfrac{4 + 3y}{5} = x$. Check your work by plugging in the value of $x$ we found.

**Solution:** We start with the original equation, $0 = 5x - 3y - 4$. Everywhere we see an $x$, we need to insert a $\dfrac{4 + 3y}{5}$. We get $0 = 5\left(\dfrac{4 + 3y}{5}\right) - 3y - 4$.

> **Kositsky's Cautions**
>
> Remember to put parentheses around the entire value of your variables for plugging in expressions. This way when you plug in $3y$ you'll remember to distribute the 3 across the whole expression for $y$ rather than just the first term. If $y = x + 1$, then $3y = 3(x + 1)$.

Rewrite the leftmost 5 as a fraction and perform the multiplication. If you look carefully at the term $\dfrac{5}{1} \cdot \dfrac{4 + 3y}{5}$, you can see that $4 + 3y$ is being divided by 5, then multiplied by 5, leaving the numerator $4 + 3y$. We now have $0 = (4 + 3y) - 3y - 4$.

---

### Practice Makes Perfect

Problem 16: Determine if
$x = 5$, $y = 20$ is a solution to

$$\frac{40+2y^2}{10} = -18x + \frac{8}{y}\sqrt{yx} + 364.$$

Subtracting from left to right, $4 + 3y - 3y = 4$, and
$4 - 4 = 0$, so $0 = (4 + 3y - 3y) - 4$.

$$0 = (4) - 4.$$

$$0 = 0.$$

Because this is a true statement our solution for $x$
was correct.

## The Least You Need to Know

◆ To solve an equation, isolate the variable on one side and the numbers on the
other side.

◆ Addition and subtraction are inverse operations and can be used to move numbers away from the variable.

◆ Division undoes multiplication and is used to strip coefficients off of variables.

◆ Whatever you do to one side of an equation, you must do to the other side of
the equation.

◆ Always check your work.

# Chapter 12

# Introduction to Algebra

## In This Chapter

- ◆ Fractions and variables
- ◆ Cross-multiplying with variables
- ◆ Solving rate-of-work problems
- ◆ Solving distance-rate-time problems

If you aren't reading this in October, do your best to get in a Halloween frame of mind. Imagine goblins and gremlins. Think about zombies and monsters. Bring to mind everything scary that haunts your nightmares. Now I want you to get all of the frightening creatures together and to command them to battle against the algebra in this chapter.

This chapter is packed to the gills with all of the algebraic techniques and word problems that populate the worst nightmares of math students. With an army of fear on your side, you can fight off the demons of algebra. Now that you're commanding things, algebra will be afraid of you, and you'll be ready to learn the rules to tackle it.

The next time someone tries to cause fear by mentioning a train that leaves New York and a train that leave Chicago and conjuring the evils of algebra, you'll be ready to do battle and win.

# Simplifying with Fractions

Ironically, despite the name "simplifying," a lot of students get freaked out by having to simplify fractions that are made up of expressions. Don't worry. Just remember the basic rules of fraction arithmetic and that the fraction bar represents division, and everything will be okay.

## Simplifying with Common Factors

You probably remember when you were working with fractions containing just numbers that you could reduce the fraction to lowest terms. (If you don't remember, check out Chapter 4.) You're going to be doing the same thing, except that some of your factors are going to be variables and not numbers. Aside from that, everything is going to be the same. Just imagine that we've invented some new numbers and they just happen to have letters for names.

**Example:** Simplify the fraction $\dfrac{24x^4}{30x^2}$.

**Solution:** Factoring the numerator and the denominator you have $\dfrac{2^3 \cdot 3 \cdot x \cdot x \cdot x \cdot x}{2 \cdot 3 \cdot 5 \cdot x \cdot x}$.

Dividing out the common factors, you're left with $\dfrac{2^2 x^2}{5}$, which equals $\dfrac{4x^2}{5}$.

This method extends to more complicated-looking problems with more than one variable in them.

**Example:** Simplify $\dfrac{120x^4 y^3}{215x^2 y^7}$.

**Solution:** Here the fraction factors as $\dfrac{2^3 \cdot 3 \cdot 5x^4 y^3}{5 \cdot 43x^2 y^7}$.

The 5 will divide out in both the numerator and the denominator, two copies of $x$ will divide out, as will three copies of $y$. This will leave you with $\dfrac{2^3 \cdot 3x^2}{43y^4}$.

Multiplying in the numerator, you get the answer $\dfrac{24x^2}{43y^4}$.

**Timely Tips**

Once you're done simplifying, each variable should be only in the numerator or only in the denominator.

**Practice Makes Perfect**

Problem 1: Simplify $\dfrac{70a^3 b^2}{150a^2 c^7}$.

Problem 2: Simplify $\dfrac{45x^3 z^4}{66x^2 y^{-2}} \cdot \dfrac{11y}{270xy^3 z^4}$.

# Positive or Negative Exponents?

Depending what's being asked when simplifying fractions of expressions, there are two ways to give the answer. You can respond with a fraction that has variables in both the numerator and denominator, all raised to positive powers. Or, you could write the expression so that it's no longer a fraction (well, maybe the coefficient would still be a fraction), and all the variables from the denominator are written with negative exponents. Neither one of these is more correct than the other, but you will probably come across a situation where the person asking the question wants the answer in a preferred form. The person asking the question always gets to pick the form for the answer.

The key to problems like this is to remember that when you change the sign of an exponent, you must move it from the numerator to the denominator (or vice versa).

♦ If you have a positive exponent in the denominator, you can change it to a negative exponent in the numerator.

♦ If you have a negative exponent in the denominator, you can change it to a positive exponent in the numerator.

♦ If you have a positive exponent in the numerator, you can change it to a negative exponent in the denominator. (There is really no reason to ever do this, but you can if you want to.)

♦ If you have a negative exponent in the numerator, you can change it to a positive exponent in the denominator.

**Example:** Convert $\dfrac{3x^7}{y^4}$ out of fraction form by using negative exponents.

**Solution:** Here you have a positive exponent in the denominator, and you want to move the factor to the numerator (to un-fraction the expression). To move this factor, you switch its sign, leaving you with $3x^7y^{-4}$.

**Example:** Write $7x^{-3}y^{-2}z^9$ with only positive exponents.

**Solution:** When we change signs of exponents, we have to move them from the numerator to the denominator. Plain numbers and plain expressions, like this one, are secretly fractions whose denominator is 1, so right now everything from this problem is in the numerator. (At the moment the denominator is invisible.)

---

**Practice Makes Perfect**

Problem 3: Write $\dfrac{24x^2}{43y^4}$ with only negative exponents.

Moving the $x$'s and the $y$'s to the denominator by changing the sign of the exponent gives you $\dfrac{7z^9}{x^3 y^2}$.

# Cross-Multiplying

It's time to bring out a powerful shortcut. Watch out for this one because if you can't remember how it works, then you need to just step away and not use it at all; if you try to use it in a case where you shouldn't, you'll mess up the entire problem. But, it's just a shortcut, so if you can't add it to your bag of tools, that's okay. I know that everyone loves a shortcut, especially people who are trying to minimize the amount of time spent on their math homework, so listen up and learn the secrets of cross-multiplying.

You might have a vague sense of familiarity with this notion. Didn't we meet cross-multiplying back in Chapter 4? Yes, yes, yes. It's a powerful and multi-talented technique that can be used to solve a wide variety of problems. Not only can it be used to determine whether two fractions are equal, but it also can be used to help solve for $x$ when your equation is made up of two equal fractions.

That's the key to it: you can use this only to solve equations when you have two equal fractions (the cross-multiply has a bit of a fixation on equal fractions).

Time for a pop quiz! In which of the following situations can you immediately use cross-multiplying to solve the equation? Don't worry that you don't know how to do it yet; that's intentional to keep you from getting ahead of yourself. Just answer the question: in which of the following situations can you use cross-multiplying to solve the equation?

(a): $a/x = b/c$

(b): $x^{-1} + {}^{17}\!/x = b/c - 6x$

(c): $a/x + 7 + z = c/x$

(d): $\dfrac{3x^4 + 7x}{\sqrt{x}}$

Did you pick (a)? That's the right answer. You can only use cross-multiplying when you have two equal fractions and nothing else in the problem. It's okay if you start out with something more complicated and simplify it to be two equal fractions—just as long as you do the simplifying before you try to cross-multiply. I certainly hope you didn't pick answer (d)! It's not even an equation since there's no equals sign!

Okay, enough of the warnings and cautions. This is only math, not running a nuclear power plant. Of course, if you plan to eventually become an engineer and work with something potentially dangerous (like nuclear power) or public safety (building bridges), then you'll have to get used to heeding warnings. In those cases, what would otherwise be a simple math error could end up in tragedy.

So how do you use cross-multiplying? It's really quite simple: you take the numerator of one fraction and multiply it by the denominator of the other and set it equal to the numerator of the second times the denominator of the first.

$$\frac{(17 - 3)}{(10 - 2)} \diagup\!\!\!\!\diagdown \frac{(3x - 4)}{(x)}$$   *Cross multiplying*

$$(17 - 3)\,(x) = (10 - 2)\,(3x - 4)$$

**Example:** Solve for $x$: $\dfrac{x}{5} = \dfrac{8}{3}$.

**Solution:** When we cross-multiply, we get the equation $3x = 40$. Solving, this is $x = \dfrac{40}{3}$.

That was easy. Why all the warnings about that? There are a few things you need to keep in mind when cross-multiplying:

♦ Remember, you must start out with two equal fractions.

♦ Sometimes if you have a plain number (like 7 or 348), you can turn it into a fraction by writing it with a denominator of 1. This can help you write things in the form two equal fractions.

♦ If $x$ is in the denominator, you need to be very, very, very, very careful. You never can divide by zero (no matter what). If there is an $x$ in the denominator, you can't have any $x$ values that would make the denominator be zero.

**Example:** Solve for $x$: $\dfrac{7}{x+5} = \dfrac{6}{x}$.

**Solution:** Since these are two equal fractions, cross-multiplying is an option. To avoid the zero in the denominator problem, you have to mention that $x$ can't be −5 and $x$ can't be 0.

Cross-multiplying, you get $6(x + 5) = 7x$.

By the distributive rule, this becomes $6x + 30 = 7x$.

Subtracting $6x$ from each side, you have $30 = x$. By tradition, we tend to write this as $x = 30$.

**Example:** Solve for $x$: $\dfrac{3}{x} = 1.6$.

**Solution:** As currently written, this isn't of the form of two equal fractions. So you can't use cross-multiplication unless you rewrite it in that form. Here it's pretty easy to do that because you can turn 1.6 into a fraction fairly easily. You can make it into an ugly fraction by giving it a denominator of 1 and write it as $\dfrac{1.6}{1}$. Some people get nervous around fractions that have decimals in them. A prettier way to turn 1.6 into a fraction is to write it as $\dfrac{16}{10}$. Doesn't matter to me how you do it, as they are both mathematically equivalent (fancy talk for "they are equal").

So now the problem is $\dfrac{3}{x} = \dfrac{16}{10}$.

Now you can cross-multiply, getting $16x = 30$. Dividing both sides through by 16, you get

$$x = \frac{30}{16} = \frac{15}{8}.$$

---

**Practice Makes Perfect**

Problem 4: Use cross-multiplying to solve $\dfrac{7a}{\frac{1}{2}+2} = \dfrac{6a}{3}$ for $a$.

---

# Working Together

You know how it goes. You've been asked to complete a task, and it will just take way too long to do it alone. So you ask someone to help you, hoping that with the two of you working together that you can each do half of the work and then you'll get done in half the time. Unfortunately, it doesn't always work that way. You've probably found yourself being helped by a total slow-poke who does less than half of the work, leaving you with more than your fair share. And yes, while it will take you less time to finish the task than if you were working alone, it's nowhere near the amount of gained leisure time that you were hoping for when you found out that you were going to be working with someone else.

There is a formula that can be used to calculate how long it will take two people working together to complete a task if you know how long it takes each of them working alone.

Let $a$ = time the first person would take, working alone.

Let $b$ = time the second person would take, working alone.

Let $T$ = total time that it will take with both people working together.

In this type of problem, $T$ will always be a smaller number than either $a$ or $b$. Even if one of the people is lazy, they will still save time by working together. When you're given two of these quantities, you can solve for the third by using the following formula:

$$\frac{1}{a} + \frac{1}{b} = \frac{1}{T}$$

**Example:** Working alone, you can clean the garage in 4 hours. Your little sister working alone can clean the garage in 6 hours. If you and your sister work together (and don't waste time tormenting each other), how long will it take the two of you to clean the garage?

**Solution:** Use the formula! $\frac{1}{4} + \frac{1}{6} = \frac{1}{T}$

You need to find a common denominator to add together the fractions on the left side of the equation. In this case, the common denominator is 12, and the equation becomes $\frac{3}{12} + \frac{2}{12} = \frac{1}{T}$. Simplifying, you have $\frac{5}{12} = \frac{1}{T}$. You have two options for solving for $T$: one is cross-multiplying, and the other is using what we know about reciprocals.

You know that the reciprocal of $T$ is $\frac{5}{12}$, so this means that the time that it takes both of you together to clean the garage will be the reciprocal of $\frac{5}{12}$, namely $\frac{12}{5} = 2.4$ hours.

**Example:** Your brother says that he can mow the lawn really, really, quickly using his special method of mowing in a spiral shape (starting with the maple tree in the middle of the front yard) instead of the usual method of going back and forth. He also has a secret way to edge the lawn, but he's not telling what that is. You can mow the lawn in 80 minutes. The two of you decide to team up on the yard work, and it only take you two 30 minutes to mow the lawn when you're working together. How long would it take your brother working alone?

*Working together at mowing the lawn.*

**Solution:** Again, the equation is going to save you. You know that you can do it in 80 minutes, so let $a$ = 80. We're trying to find how long it takes your brother to mow the lawn, so let that be $b$. And you know that it takes the two of you 30 minutes working together.

$$^1/_{80} + {^1/_b} = {^1/_{30}}$$

To solve this for $b$ you'll subtract $^1/_{80}$ from each side of the equation $^1/_b = {^1/_{30}} - {^1/_{80}}$. This requires a common denominator; the common denominator for 30 and 80 will be 240.

$$^1/_b = {^8/_{240}} - {^3/_{240}}$$

$$^1/_b = {^5/_{240}}$$

$$^1/_b = {^1/_{48}}$$

Since the reciprocal of $b$ is $^1/_{48}$, $b$ must equal 48. Your brother with his secret shortcuts could mow the lawn in 48 minutes.

> ### Practice Makes Perfect
>
> Problem 5: Trillian and Arthur need to pump water from a nearby stream into their pool. If Trillian could fill the pool by herself in 300 minutes and Arthur could fill it by himself in 900 minutes, how long will it take them working together?

# Planes, Trains, and Automobiles

When people make fun of algebra, one of the typical things that they do is start to recite a make-believe word problem where one train leaves Chicago and another leaves New York. At this point the problem starts to devolve with all sorts of scary and extraneous details into a morass of numbers that no one has any chance of understanding—let alone solving. Just like there's that saying that stereotypes come from somewhere, there is a reason why people use this type of problem to make fun of algebra. One of the types of problems that comes up in just about every algebra class is the distance-rate-time kind of problems.

The distance-rate-time problem is so classic that it is previewed in pre-algebra, studied to death in algebra, and resurrected in pre-calc and in calculus. You can't escape these problems. So, take a seat, and enjoy the ride.

All of these problems are based on what is sometimes referred to by students as the "dirt equation." In words, this equation says that distance equals rate times time. Symbolically, it's $d = rt$, which looks sort of like the word "dirt."

The units in these types of problems need to make sense together. If your distance is measured in meters and your time is measured in seconds, then it's absolutely essential that the time is measured in meters per second. When distance is in miles and time is in hours, then rate has to be in miles per hour. If you have incompatible units, then before starting the problem you have to convert one of them so that they all work together.

The simplest type of these problems is when you're given two of the pieces of information and are asked to find the third. You could be given distance and time and be asked to find the rate. Or, maybe you'll be given distance and rate and then solve for time. If you're really lucky, you'll be given rate and time and then multiply them together to get distance.

**Example:** You travel 221 miles at 65 miles per hour. How long were you traveling?

**Solution:** The equation is $d = rt$. Distance is 221 and rate is 65, so the equation becomes $221 = 65t$. Dividing both sides of the equation by 65, you get $t = 3.4$ hours.

Unfortunately, things get more complicated. You'll have two trains. Or maybe someone will travel somewhere and back but the problem will leave out what you might think are important details. Maybe there are multiple destinations.

The best way to handle the overload of travel information is by organizing what you know in a chart. You'll collect what you know about the distance, rate, and time of two different trips. Sometimes they'll be two parts of the same trip.

|        | *distance* | *rate* | *time* |
|--------|------------|--------|--------|
| Trip 1 |            |        |        |
| Trip 2 |            |        |        |

In this chart you will fill in whatever you are told about distance, rate, and time. Sometimes the information will be explicitly given to you, like a car might be traveling at 65 miles per hour. Sometimes you might need to figure out what variables to introduce in order to describe distance, rate, or time. Finally, you use the $d = rt$ equation to come up with information about what's lacking.

**Example:** David decides to drive from St. Louis, Mo., to Memphis, Tenn., to visit Graceland. On his trip to Graceland, he travels at an average speed of 65 miles per hour, and it takes him 4.8 hours to make the trip. Because of construction on the road, on his way home, he travels at an average speed of 60 miles per hour. How much longer did his return journey take?

**Solution:** It's true, you don't need to use the chart method to solve this problem. Use it anyway. There are other problems where the chart will be super-helpful, and you want to be ready for those.

You start out by filling in what you know. In this problem the first trip is the journey from St. Louis to Memphis, and the second trip is the way back. You can fill in the information that you know: both rates and one of the times.

|        | distance | rate | time |
|--------|----------|------|------|
| Trip 1 |          | 65   | 4.8  |
| Trip 2 |          | 60   |      |

You can find the distance from St. Louis to Memphis by using the $d = rt$ equation, and $d = 65 \times 4.8 = 312$ miles. The problem assumes that David takes the same route back to St. Louis, so both distances will be the same, and you can fill in 312 for both distances in the chart. The only unknown left in the problem is the return time. You can call that $t$.

|        | distance | rate | time |
|--------|----------|------|------|
| Trip 1 | 312      | 65   | 4.8  |
| Trip 2 | 312      | 60   | $t$  |

To find the return time, you use the $d = rt$ equation for the return trip: $312 = 60t$.

Dividing both sides by 60, you get $t = 5.2$.

The problem asked how much longer the return trip was. Because it took David 4.8 hours to drive to Memphis and 5.2 hours to drive back, the difference is $5.2 - 4.8 = 0.4$ hours, which is 24 minutes.

Let's now consider the classic problem of when two cars or trains leave different cities and are heading toward each other, destined for a fiery collision unless you can solve the equation. Feel free to take a break now and enjoy the impending fireworks.

In these problems, the trains start out at the same time, traveling at different rates, and are heading toward each other. In this type of situation, the problem almost always will be solved in the following way.

The key to this type of problem is that while they are heading in different directions and are traveling different distances, they will both have the same time. They both start at the same time. They both end up in the catastrophic (and pretty) fireball formed when they collide at the same time. Time is the same. This will be the connection between the two trips. While much less dramatic, you also can use this same method to solve problems where the two trains leave the same city heading in separate directions. In this case, instead of trains, we'll look at cars.

**Example:** Two cars leave Springfield, Mass. One heads east on the Massachusetts Turnpike at a speed of 60 miles per hour, and the other heads west at a speed of 65 miles per hour. After how many hours will they be 297 miles apart?

**Solution:** You start by filling in the chart with the easiest information to deal with. In this case, it is most definitely the rates.

|        | *distance* | *rate* | *time* |
|--------|-----------|--------|--------|
| Trip 1 |           | 60     |        |
| Trip 2 |           | 65     |        |

Yes, easy so far, but so what? The next step is to introduce a variable. In these problems because the time is the same for both cars (trains, racing snails, whatever), your variable will be time. Also, the original problem is asking you to solve for time, so having time as a variable is a good thing for that. Add time to the chart.

|        | *distance* | *rate* | *time* |
|--------|-----------|--------|--------|
| Trip 1 |           | 60     | $t$    |
| Trip 2 |           | 65     | $t$    |

Now you need to fill in distance. I know that you are very, very, very, very tempted to take that 297 miles and put it in the chart. Don't do it! The 297 miles can't be assigned to either the eastbound car or the westbound car, so there is no room for it in the chart. You can't put it anywhere; it doesn't go in the chart. Do not put it in the chart!

So how do we fill in distance? We do it with the $d = rt$ equation. You might be skeptical. You might wonder how it will help to have more expressiony mathematical gobbledygook filled with variables when there is a nice plain number floating around unused. Just trust that this will work. When you multiply rate by time, you get $60t$ and $65t$.

|        | *distance* | *rate* | *time* |
|--------|------------|--------|--------|
| Trip 1 | $60t$      | 60     | $t$    |
| Trip 2 | $65t$      | 65     | $t$    |

Now, you're probably wondering so what. Think back to what you're being asked to do. The original problem asks you to find the time at which the cars are 297 miles apart. Now you're ready to use the 297. (It's all about doing the right thing at the right time.) You're going to set up an equation where the total distance traveled is 297 miles, and where the variable is $t$. Fortunately, you can do that without much trouble. You know that the first car traveled a distance of $60t$ and the second car traveled a distance of $65t$ (this is why having a chart is so helpful), and these two distances add up to 297 miles. Turning that into an equation, you get $60t + 65t = 297$.

Now you have an equation that you can solve.

Simplifying, you get $125t = 297$.

Dividing both sides by 125, you have $t = 2.38$. Therefore, the total travel time is 2.38 hours.

**Kositsky's Cautions**

Don't turn 2.2 hours into two hours and 2 minutes or two hours and 20 minutes.

So what happened to our giant fireball of pyrotechnics and car crashes and explosions and special effects? We started with the promise of excitement and ended up with a few imported sedans driving around in western Massachusetts. How do you solve problems where the vehicles start in different places and then meet in the same place (whether they arrive safely or not)?

Imagine taking the previous example of the cars in Massachusetts and filming it. Now imagine running the film backward. Ignore the part about the cars driving backward—this is Hollywood. With the film running backwards, the cars start out in different parts of the state and then meet in Springfield. They're still going at the same speed. They're still covering a combined distance of 297 miles. And they will still do it in 2.2 hours. Whether the cars start in the same place and head away from each other or whether they start in different places and are heading toward each other, you will do the problem the exact same way.

**Example:** You are meeting your grandmother by the side of the highway for a secret exchange of gifts (because your mom is sick of your grandma spoiling you). You start out in cities that are 180 miles apart, and you drive toward each other. You drive at a speed of 65 miles per hour, and your grandmother drives at a speed of 55 miles per hour. When will you meet? How many miles will you each have traveled? Why can't more grandmothers spoil their grandchildren?

**Solution:** You need to find the time that you've both traveled. From that point everything else will work itself out.

Once again, set up the chart and fill in what you know:

|        | distance | rate | time |
|--------|----------|------|------|
| Trip 1 |          | 65   |      |
| Trip 2 |          | 55   |      |

Again, you will have the variable be time:

|        | distance | rate | time |
|--------|----------|------|------|
| Trip 1 |          | 65   | $t$  |
| Trip 2 |          | 55   | $t$  |

And you can find expressions for the distances by using the $d = rt$ formula:

|        | distance | rate | time |
|--------|----------|------|------|
| Trip 1 | $65t$    | 65   | $t$  |
| Trip 2 | $55t$    | 55   | $t$  |

And you know your combined distances are 180 miles, so $65t + 55t = 180$.

Simplifying, $120t = 180$. Solving, $t = 1.5$ hours.

This problem also asked you to find the distances that were traveled. Once you know the time, you can calculate them from the $d = rt$ formula. Your grandmother traveled $55 \times 1.5 = 82.5$ miles and you traveled $65 \times 1.5 = 97.5$ miles.

---

| **Practice Makes Perfect** |
|---|
| Problem 6: A train leaves Los Angeles headed for San Diego at 8 a.m. at a constant speed of 50 miles per hour. At the same time, Dagny heads north along the train tracks at 80 miles per hour to stop the train before it gets to Oceanside. Los Angeles and San Diego are 120 miles apart and Oceanside is 50 miles away from Los Angeles. Will Dagny meet the train in time to stop it? |

## The Least You Need to Know

♦ In a fraction you can generally move around $x$'s as if they were normal numbers and not variables.

♦ When you cross-multiply, make sure you multiply the entire numerator by the entire opposite denominator on both sides of the equation.

♦ Use the equation $\frac{1}{a} + \frac{1}{b} = \frac{1}{T}$ to find how long $T$ it will take two people who normally would finish in $a$ and $b$ to do the task together.

♦ $d = rt$ is the fundamental equation to almost any problem involving distance, rate, and time.

# Part 5

**Ancient Foundations: Geometry**

A long time ago, a guy named Eratosthenes measured the distance around the Earth by knowing little more than the distance between two cities and a little bit about the rays of the sun. That was the birth of geometry. Since then geometry has grown up into a field of mathematics that is brimming with shapes, figures, and equations. In this part, we'll issue you a field guide to triangles, instructions on how to use a protractor, and more formulas than you can shake a stick at.

*"Not so dismissive of surface area now, are ya?"*

# Chapter **13**

# Shapes on a Plane

## In This Chapter

- ◆ Plotting points
- ◆ The coordinate plane
- ◆ Drawing graphs
- ◆ Rigid motions in the plane

One of the oldest branches of mathematics is geometry. It's probably not as old as counting, but it's still pretty old. Geometry is named for words meaning "Earth measure," and one of the oldest known examples of geometry in action is when Eratosthenes was able to measure the circumference of the Earth by knowing a little bit about the sun's rays, angles, and the distance between two cities.

But nowadays it's much easier to search for the circumference of the Earth on the Internet, and most geometry is done on paper in schools. Less grandiose, but much more practical and sane.

You are probably asking what a chapter on geometry is doing in a book on pre-algebra. In this chapter you start learning about the connections between the two.

# Plotting Points

Drawing pictures that come from algebra is really just a grown-up version of connect the dots. The main differences are that you need to put the dots on the page, and you aren't explicitly told which order to connect them in. While this might not seem like it's much of an analogy, it holds true. Each time you have to do one of these problems, you'll be told where the dots come from and where to draw them on the paper. Later, you'll learn about connecting them.

## Remembering the Number Line—Cartesian Coordinates

You remember the number line, right? It was easy. If I told you a number like 7 or 3.2 or whatever, you could plot it on the number line. You're going to do the same thing, except instead of restricting yourself to the number line, you'll have access to the entire page. You'll gain this freedom by adding a second number line into the mix—in a specific way.

*The points 7 and 3.2 plotted on the number line.*

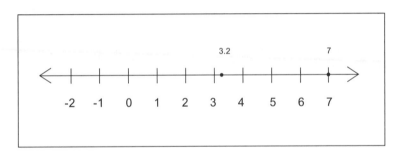

Any flat surface that goes on forever is called a plane; typically we look at only a small part of a plane that can be drawn on a piece of paper or on the board at the front of a classroom. When we talk about the Cartesian plane, we also are including a grid system with which we can describe where any point on the plane is located.

We take two number lines and put them on the page in the following way: one of the number lines runs horizontally and the other number line runs vertically. They cross at the place marked zero on each of them. On the horizontal number line, the positive numbers are to the right, and the negative numbers are to the left. On the vertical number line, the positive numbers increase as you head up, and the negative numbers descend as you head down. All of these things have names.

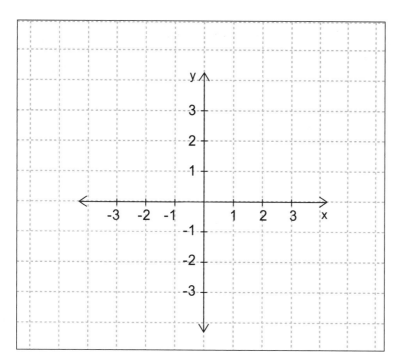

*The Cartesian plane.*

◆ The horizontal number line is called the *x*-axis. If your math teacher is from the nineteenth century, you might hear it called the "abscissa." Your teacher may also be a vampire.

◆ The vertical number line is called the *y*-axis. The archaic name for this that no one (except vampires) uses anymore is the "ordinate."

◆ The *x*- and *y*-axes together often are called the coordinate axes.

◆ The place where the *x*-axis meets the *y*-axis, and is numbered 0 on both of them, is called the origin.

◆ This entire set-up, the plane together with the axes providing a grid system for numbering it, is called the Cartesian plane or the coordinate plane.

**Amy's Answers**

In case you haven't seen the word "axes" used in this context before, it's the plural of "axis." I remember wondering in high school why math teachers were so obsessed with axes. I thought it was to keep us students in line.

If your math teacher talks about the abscissa and the ordinate, they really need to get up to date because no one talks like that anymore, unless they're showing off. And, yes, the names $x$-axis and $y$-axis do refer to the same $x$ and $y$ that you see in expressions and equations.

*How to plot the point (2,3) on the Cartesian plane.*

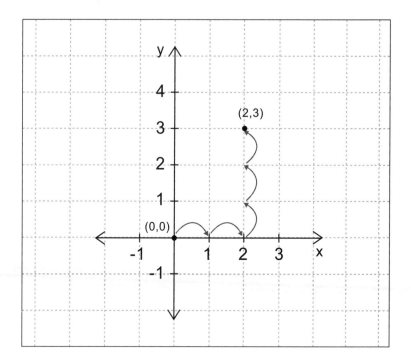

Because you have two number lines, you will have two numbers to describe each point. Points will be given as ordered pairs of numbers written in parentheses and will look something like (2,3). The first number will tell you how far to travel along the $x$-axis, and the second number will tell you how far to travel up the $y$-axis. These numbers are called the $x$-coordinate and the $y$-coordinate of the point. Together, they're called the *coordinates* of a point, and we write them in the form $(x, y)$. The ordering of coordinates will be important for the next year or two of your mathematical career, so memorize it well. $x$ before $y$. Horizontal before vertical. Left-right before up-down.

The coordinates of a point can be any number. We tend to use numbers that are fairly modest in size because plotting large numbers typically requires using large paper.

**Kositsky's Cautions** _____

It's important to note that points are ordered pairs. It matters which number comes first, and remember that this is the *x*-coordinate. The second is always the *y*-coordinate.

Also, the parentheses used in writing a point have nothing at all to do with the parentheses that are used in multiplying or in the order of operations. It was a choice between re-using symbols for different purposes or else designing a crazy gibberish that looked like it was written in a Martian language. Mathematicians opted for re-using symbols as we haven't found any decent Martian languages.

**Example:** Plot the points (1,2), (–4,0.5), (3,0), (–2,–2).

**Solution:** First we need to set up our graph. Draw a horizontal line and label it the *x*-axis. Don't forget to put the two arrows on the end! Then draw a vertical line and label it the *y*-axis. Unless you can tell where on the graph you'll end up drawing, you should probably place the *y*-axis in the center of the *x*-axis. Make even tick marks going along both axes. Usually, you'll want (0,0) to be where the *x*- and *y*-axes intersect, the first tick mark to the right of (0,0) to be 1 and the first tick mark above (0,0) to be 1. This means you move a distance of 1 when you move one tick mark vertically or horizontally. While it's possible to have the distance between adjacent tick marks be any quantity you want (such as 2 or 0.7), we'll have the distance be 1 in this problem.

Let's think of this as four simpler problems, one for each point we're going to plot. First we want to plot (1,2). Since each tick mark represents a distance of 1, we want to start with our pencil at (0,0), move it one tick mark to the right, then 2 tick marks up. This is the point (1,2), and we label it as such. Repeating these steps for each pair of points, we end up with the following figure.

There are some things that you need to keep track of when you're plotting points (or anything else) on the coordinate plane. Most important, there are a lot of things that need to be labeled. Once you are a professional mathematician, you can get away with leaving off some of the labels, but until then, you will need to label all of these things all of the time (unless you are explicitly told otherwise). These are the sorts of things that teachers like to take points off for. Don't lose points for things that are this easy; save them for the things that are really difficult.

*(1,2), (–4,0.5), (3,0), and (–2,–2) plotted on the Cartesian plane.*

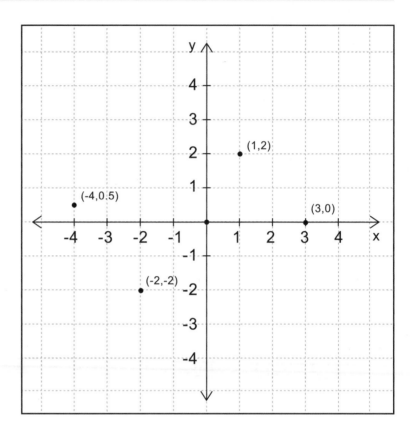

### Amy's Answers

Don't have any graph paper on hand? You have three main options.

1. Make it by hand. If you're either really patient with a ruler or really good at drawing parallel lines freehand, this works.

2. For the rest of us, most stores that have school supplies have graph paper.

3. My favorite option, so I don't have to leave my room, is downloading graph paper off the Internet and printing it. Just Google "graph paper" and something usable should come up.

The first thing is that the *x*-axis needs to be labeled *x* and the *y*-axis needs to be labeled *y*. This may seem obvious, as they never seem to move. Still, you need to label them every time. Sometimes you will be doing problems in which the variables have names other than *x* and *y*, and you would need to label them appropriately. The *x*- and *y*-axes also both need to have arrowheads on both ends. This tells the reader of your graph that the axes go on forever in both directions, just like the number line.

The other thing that needs to be labeled is the scale on your axes. Does one tick mark mean that you're at the number 1? Or have you smooshed the scale down so that each tick mark has numbers that you find by counting by twos? Or is it so zoomed in that each mark on your axes is a tiny fraction or decimal? You need to tell the viewer of your graph what the marks on your axes represent. Whatever your scale happens to be, it's very important that the distance between consecutive tick marks along the $x$-axis is the same no matter where you are. The same is true for tick marks along the $y$-axis.

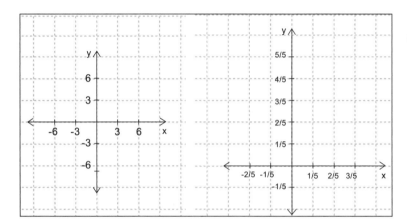

*Two Cartesian planes with different scales on the axes.*

Problem 1: Plot the points (–2,1), (2,0), (0,–1), (2,4) on a Cartesian plane with the distance between two tick marks equal to $\frac{1}{4}$ on both axes.

Problem 2: Plot the points (3,5), (0,2), (–4,–3), (0,4) on a Cartesian plane with the distance between two tick marks equal to 2 on both axes.

# Graphing Equations

Not every equation you meet is waiting for you to solve it and give one specific number as a right answer. If there is only one letter in the equation, then you have a shot at solving it and getting one number as a solution. Sometimes in algebra class, there might be two or more solutions. Now, when you introduce a second letter, like $y$, into an equation and find yourself with something like $y = x + 1$, you'll find yourself with infinitely many solutions. Seriously. More than a zillion. Infinitely many. There is no way that you can list all those numbers on a sheet of paper in your lifetime, so instead you summarize them all by drawing a picture.

What does a solution to an equation with two variables in it look like? It is a pair of numbers: one for $x$ and one for $y$. For example, one of the many solutions to $y = x + 1$ is the pair $x = 4$ and $y = 5$. Can you see why this satisfies the equation? If you plug in 4 for $x$ and 5 for $y$, you get $5 = 4 + 1$, which is true.

Traditionally, we won't write this solution out the long way as $x = 4$ and $y = 5$. Instead, we write this as an ordered pair with the $x$ value first and the $y$ value second. The solution $x = 4$ and $y = 5$ would be written $(4, 5)$. Does this look familiar? This is a point that can be graphed on the Cartesian plane. This is the main idea behind graphing equations. You try to draw a picture that represents each and every one of the infinitely many solutions to the equation.

Making a list of a paltry few of the infinitely many solutions to the equation is a process known as making a table of values. To make a table of values, you select values for $x$ (you get to choose which ones), then you plug them in and find the corresponding values of $y$. Typically, you choose $x$ values that are integers near 0; numbers like −2, −1, 0, 1, and 2 are fairly typical. You might be asked to use different numbers. You might have the freedom to try some others out. Most of the time these $x$ values will work just fine.

To graph the equation, carry out the following steps.

1. Make a table of values.

2. Make a list of ordered pairs from your table of values.

3. Plot the points corresponding to those ordered pairs on the Cartesian plane.

4. Connect the points. If it's a straight line, try to draw it as straight as possible. If there's a curve, try to make it smooth.

**Example:** Graph the equation $y = x + 1$.

**Solution:** Start by making a table of values. Choose some values for $x$ (I'm going to be traditional and pick −2, −1, 0, 1, and 2), plug them in for $x$ and calculate $y$ using the formula $y = x + 1$.

| $x$ | $y$ |
|-----|-----|
| −2 | −1 |
| −1 | 0 |
| 0 | 1 |
| 1 | 2 |
| 2 | 3 |

From this chart, we now have the five points (–2, –1), (–1, 0), (0, 1), (1, 2), and (2, 3).

Plot these five points on the Cartesian plane. They line up in a line. Draw a line connecting them. Depending on the whims of your teacher, you may need to put arrowheads at both ends of the line to show that the line goes on forever in both directions and contains all of the infinite solutions to the equation. You also should write the equation of the line (that $y = x + 1$ thing) next to the line so it's really clear what you are graphing.

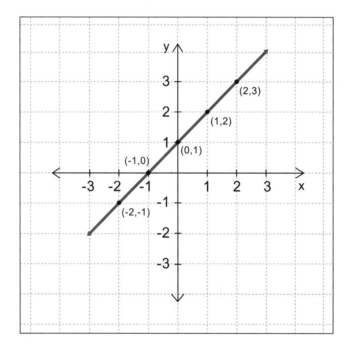

*The line $y = x + 1$ plotted on the coordinate plane.*

**Example:** Graph the equation $y = x^2 – 1$.

**Solution:** Start by making a table of values. Once again, choosing $x$ values of –2, –1, 0, 1, and 2 will be a good choice. Plug each of these numbers in for $x$ and calculate the corresponding value for $y$.

| $x$ | $y$ |
| --- | --- |
| –2 | 3 |
| –1 | 0 |
| 0 | –1 |
| 1 | 0 |
| 2 | 3 |

We get the five points (–2, 3), (–1, 0), (0, –1), (1, 0), and (2, 3).

Plot these five points on the Cartesian plane. They curve around into a bit of a U-shape. Don't connect them with straight lines like you would in a non-mathematical connect-the-dots. Instead draw a smooth curve that swoops down to graze each of the points. This is not exactly what the equation $y = x^2 - 1$ looks like, because we haven't graphed every point $(x,y)$ where $y = x^2 - 1$ is true. It turns out there are an infinite number of those points, so graphing them all would take a long time. However, the plot-lots-of-points-and-curvedly-connect-the-dots is close enough for most purposes.

### Practice Makes Perfect

Problem 3: Graph 5 points of the equation $y = 2x + 1$.

Problem 4: Graph 10 points of the equation $y = x^2 + x$.

*An approximation of the curve $y = x^2 - 1$.*

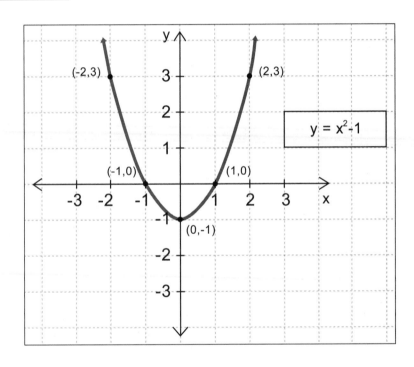

$y = x^2 - 1$

# Measuring Distance

To find the distance between two points, you use a formula called the distance formula. Let's say that you have two points called $(x_1, y_1)$ and $(x_2, y_2)$. If you're asked to calculate the distance between those two points, you use the following formula:

$$d = \sqrt{(x_2 - x_1)^2 + (y_2 - y_1)^2}$$

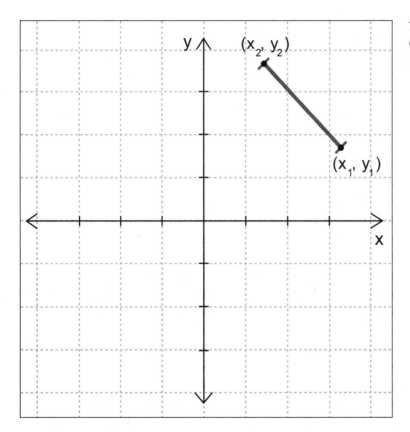

*The distance between points $(x_1, y_1)$ and $(x_2, y_2)$.*

**Timely Tips**

If you move diagonally, the line you follow makes a right triangle (a triangle with a 90-degree angle) with the coordinate grid. This is no accident! Right triangles are much nicer than other shapes. You'll see this formula for *d* come up again in Chapter 15 as the Pythagorean Theorem for right triangles.

**Example:** Find the distance between the points (2, –4) and (–5, 1).

**Solution:** While the calculation is unpleasant, it isn't difficult. Just put the numbers into the formula. In this case $x_1 = 2$, $y_1 = -4$, $x_2 = -5$, and $y_2 = 1$. Plugging in, you get $d = \sqrt{\left(-5-2\right)^2 + \left(1-\left(-4\right)\right)^2}$ . Now you just need to be careful to work with the rules of integers and the order of operations and then deal with the radical at the end.

Simplifying, you get $d = \sqrt{\left(-7\right)^2 + \left(5\right)^2}$ , which is $\sqrt{49+25} = \sqrt{74}$ . Because the radical can't be simplified, the answer is $\sqrt{74}$ . If you're not sure if your answer makes sense, evaluate your answer on a calculator and write down the answer. Then graph the points on a coordinate plane. Measure the distance between the points with a ruler, see how far this is by placing the ruler immediately below the $x$-axis, and see if that answer is about the same as what the calculator gave you. This might not work if you draw your graph paper by hand; it certainly doesn't when I draw it by hand!

*The points (2, –4) and (–5, 1) in the Cartesian plane with their distance.*

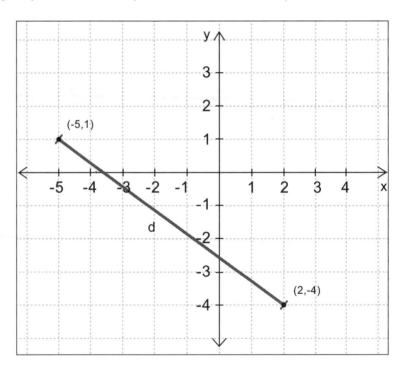

Another formula you might see is the midpoint formula. This formula lets you find the point that is halfway between two given points. Once again, say that the two points are called $(x_1, y_1)$ and $(x_2, y_2)$. The midpoint of the two points has $x$-coordinate $\frac{x_1 + x_2}{2}$ and $y$-coordinate $\frac{y_1 + y_2}{2}$.

**Example:** Find the midpoint of (–2, –1) and (4, –3).

**Solution:** The $x$-coordinate is $\frac{-2+4}{2} = 1$ and the $y$-coordinate is $\frac{-1-3}{2} = -2$. So the midpoint between these two points is (1, –2).

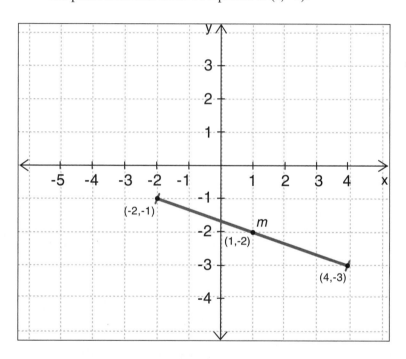

*The points (–2, –1) and (4, –3) with the midpoint between them.*

# Four Types of Rigid Symmetry

When you've heard about symmetry in the past, it was probably restricted to mirror symmetry. Mathematicians have extended the definition of symmetry to include four types of symmetry that can be applied to designs in a plane. If you're looking at an example with the coordinate plane, you can calculate the location of the symmetrical images. But if you're just on unmarked paper, you'll merely observe the attractive and symmetrical designs.

To a mathematician, symmetry has a fairly specific meaning. Imagine that you have your design on a sheet of paper and another exact copy of your design on a clear piece of plastic. Pick up the plastic, move it around, spin it about, flip it over. Do anything that you want, just as long as you don't stretch, break, or distort the plastic. If you can put it back down somewhere on the paper and the image on the page matches up with the image on your plastic, then the image is said to have symmetry.

This analogy of imagining an image and a copy of it on plastic is going to be useful throughout this section. Don't discard it from your mind yet. The four types of symmetry come from what are called rigid motions of the plane. This is just a fancy name for what you've just imagined: pretending that the image is on a clear piece of plastic and moving it around.

## Slip Sliding Away

The first type of symmetry is based on a motion called translation. With translation, you are sliding the image to a different place. Imagining our plastic, this would be like sliding the plastic to a different part of the page but without rotating it or flipping it over. You're just sliding it in a straight line.

To apply a translation to an image drawn in the coordinate plane, you'll be given a number to be added to every $x$-coordinate and a number to be added to every $y$-coordinate.

**Example:** A rectangle in the plane has coordinates (1, 2), (1, 5), (3, 2), and (3, 5). Apply the translation that adds 4 to every $x$-coordinate and subtracts 2 from every $y$-coordinate.

**Solution:** (1+4, 2–2) = (5, 0)

(1+4, 5–2) = (5, 3)

(3+4, 2–2) = (7, 0)

(3+4, 5–2) = (7, 3)

The new rectangle has coordinates (5,0), (5, 3), (7, 0), and (7, 3).

We say that an image has translation symmetry if you can get it to match up again after you slide it. What does that mean? If you had a piece of graph paper that went on forever, and then you photocopied it onto a piece of overhead transparency plastic, you could line them up so that the lines match up. If you slide the overhead transparency to the side, you can make the lines match up again. This is what we mean by translational symmetry.

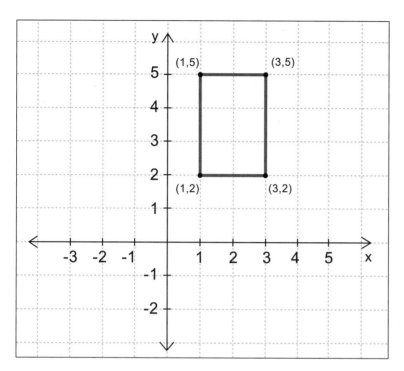

*The original rectangle.*

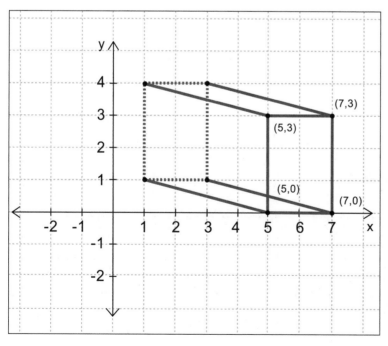

*The original rectangle grayed out and the translated rectangle.*

# Turn, Turn, Turn

The next rigid motion of the plane is a rotation. Once again, we're imagining having an image on a page and a copy of it on a piece of overhead transparency. To think about rotational motion, you want to imagine a paper fastener that pokes through both layers and holds them together at one point—but that will allow the top layer to rotate. The act of rotating the top layer is, unsurprisingly, the motion called a rotation.

It is possible to calculate what happens to a point in the coordinate plane after it is rotated. There's a good news/bad news thing going on with this, though. We'd love to tell you about it for the sake of completeness. However, it requires trigonometry, something that you probably aren't going to study for a while. If you're feeling like your life is incomplete because you can't rotate points in the plane, wait until you take trigonometry and then ask your teacher or someone in the know. Teachers tend to like it when you ask that type of question; it shows that you care about math.

An image is said to have rotational symmetry if after you rotate it, the image still lines up. Some common objects with rotational symmetry are hubcaps and playing cards.

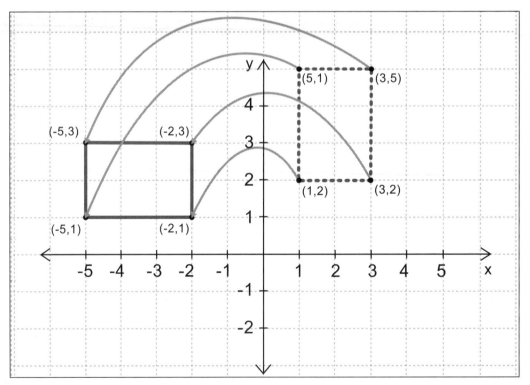

*An example of rotating a rectangle 90° counterclockwise about the origin (the origin is another name for (0,0)).*

## Into the Looking Glass

The symmetry that you're most familiar with is mirror symmetry, and it is related to the rigid motion called a reflection. Following our analogy of having an image on the page and a copy of the image on a piece of overhead transparency, the way that we implement a reflection is to pick up the plastic and flip it over. There's a bit more to it: your reflection has to be across a line that functions as the mirror. Wherever the mirror line is on the original paper-and-transparency combination, it needs to be in the same place after the motion, too.

Breaking from our analogy for a minute, you also can imagine a reflection as what happens if you draw with a wet-ink marker and then fold the paper and rub the two sides together. You get a mirror image of the drawing on the other side of the fold. Either of these should give you the same result, and they are both considered reflections.

To take an image in the coordinate plane and reflect it across a line requires little more than subtraction—as long as the line is vertical or horizontal. It's especially easy if the line is the *x*-axis or the *y*-axis. To reflect points across the *x*-axis, you change the signs of all the *y*-coordinates. To reflect points across the *y*-axis, you change the signs of all the *x*-coordinates.

**Example:** Reflect the points (1, 3) and (–4, –7) across the *x*-axis.

**Solution:** Just change the signs on the *y*-coordinates: (1, –3) and (–4, 7).

An image has reflection symmetry if it is the same after you take a mirror image.

*An example of two points being flipped across the x-axis via a reflection.*

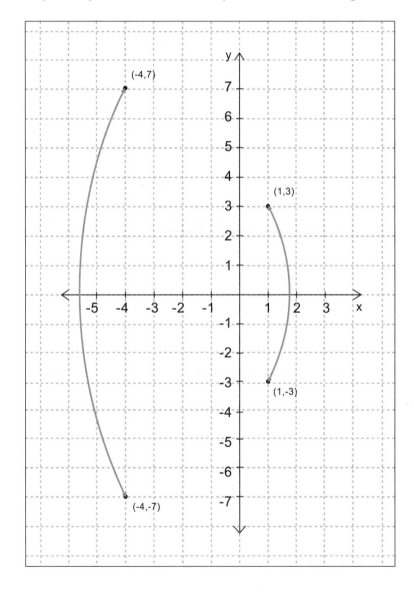

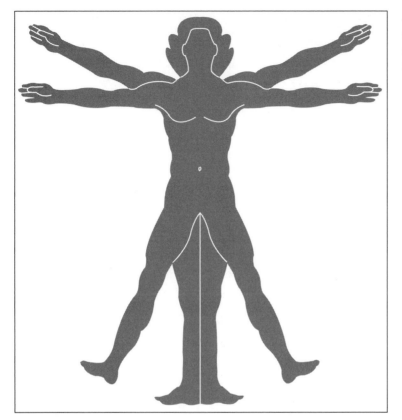

*An example of a figure with reflection symmetry. Special thanks to our friend Leo for the inspiration.*

## Footprints in the Sand

The final rigid motion is always listed third or fourth on a list of the rigid motions. This is because it's not really new and different in its own right, it's a combination of a translation and a reflection. Its name is glide reflection.

Back to imagining our image drawn on paper and on a piece of overhead transparency. To perform a glide reflection, first you flip over the transparency to do the mirroring, then you slide the transparency along the mirror line. Those two motions taken together combine to form a glide reflection.

The most common example of a glide reflection is footprints. A left footprint is the mirror image of a right footprint, but it's not directly across from the right footprint (unless you're jumping with your feet together). The footprints are staggered.

*Animal tracks in sand or mud are an example of a glide reflection.*

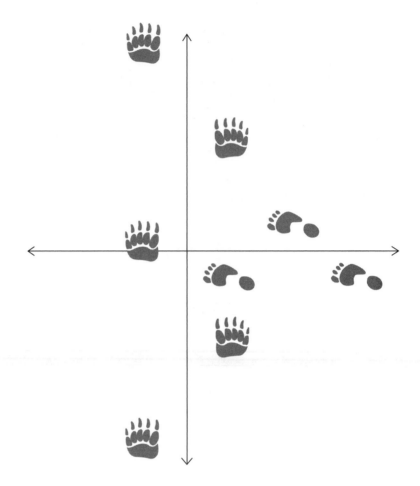

## The Least You Need to Know

- Plotting points on the coordinate plane is just like marking points on the number line—you just need to find two numbers for each point!

- Vectors can be added numerically or graphically.

- The formula to find the distance between two points on the plane is
$$d = \sqrt{\left(x_2 - x_1\right)^2 + \left(y_2 - y_1\right)^2}.$$

- There are four types of symmetry: translation, reflection, rotation, and glide reflection.

# Chapter 14

# Angles in the Outfield

## In This Chapter

♦ How to measure angles

♦ Different types of angles

♦ Special pairs of angles

♦ Parallel and perpendicular lines

Why study angles? If you're running a medieval siege operation, and you're hoping to fling rocks over the wall of the castle, you'll need to calculate the angle at which to pitch your catapult. Maybe you're into virtual warfare and want to write video game software? The same principles will apply.

If you're more peacefully inclined, angles will come up in construction, where roads meet in a housing development, and in the design of craft projects. They play a role in determining how much fun you can have sliding down a snowy bank in a sled or flying down a chute in a water park. Skateboarders and snowboarders describe some of their tricks in terms of the degree measure of the angles that they spin around. Angles will haunt you in physics class when you have objects sliding down inclined planes. In your more immediate future, you're going to need to know about angles to study triangles and polygons.

In this chapter, you learn the vocabulary associated with angles so that you can talk about them and describe them. You learn about how to measure them and how to apply those measurements.

# Measuring Angles

Roughly speaking, an angle is what happens when two lines meet. As you can tell by now, mathematicians never stick to approximate language, and everything must have a name. Speaking precisely, to talk about angles we first need to introduce the concept of a ray. A ray is a line that begins at a point and then heads out forever in one direction. A great example of this is the sun's rays. Each ray of sunlight starts at a point (the sun) and then heads out forever into space (unless it hits something). Similarly with ray guns. The ray starts at the ray gun and then heads forever along a straight line. Cosmic rays start somewhere out in space and head out in a straight line into space.

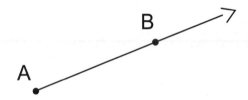

But back to mathematics. When two rays share the same endpoint, they create an angle. This point is called the vertex of the angle. Vertex is just a fancy math word that means "corner" in almost every situation. And when we talk about angles, we tend to give names to them by putting letters on the rays. This means that angles tend to have three-letter names, much like many television networks.

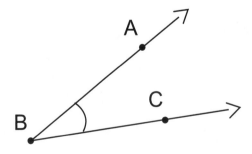

By tradition, when we give the name of an angle, we give the letter of the vertex in the middle. You need to do it this way; otherwise, you will almost certainly lose points. This isn't the sort of tradition that you can ignore. It goes deep into the mathematical culture.

**Example:** Name the following angle:

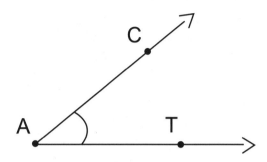

**Solution:** This angle is named CAT. Since the vertex is A, it needs to go in the middle position. You would be just as correct if you called the angle TAC. If we planned ahead, we might have been cleverer and labeled the points in a way so that the name spelled a word in both directions. Of course, the name of an angle doesn't need to spell a word, unless you speak a language that has words like PQR and LMN.

When we're being very technical and precise, there is even more jargon attached to the rays that make up an angle. In this way, math is much like field biology. Biologists have given names to all sorts of plants and animals and more names to all of the parts of the plants and animals and still more names to the parts of the parts. The organs have names and are made out of cells with names; the cells are made out of organelles (which all have names); the parts of the cell are made out of molecules, which also have names, names all the way down to the most basic structures.

Mathematicians have done much the same thing with the mathematical objects. The main difference is that you can do math without having to go outside or work with chemicals and test tubes or worry about being bitten by something. But when it comes to this type of situation, it's the same whether you're doing math or biology: you need to learn the names of all the parts.

Extending this analogy to rays and angles, we can give more names to the rays that we've been building angles out of. Sometimes we specify one of the rays of an angle as the initial side and the other ray of the angle as the terminal side. This gives us information about which way the angle opens. The angle will start at the initial side and open up toward the terminal side. You might want to imagine the blades of a pair of scissors or the two sides of a compass (the pointy type that you use in geometry, not the round type that you use in orienteering). As the two sides move away from each other, the angle opens. If you hold one of the sides fixed, that's the initial side. The one that moves is the terminal side. Again, by mathematical convention, we like the initial side to be oriented horizontally with the arrow of the ray pointing to the right and for the angle to open in a counterclockwise motion.

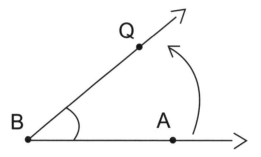

There are two main standards in use for measuring angles. The only one you are going to see in pre-algebra is the degree measurement of an angle. A degree is a measurement of how far you have rotated. If you stand in one place and spin around in a full circle, you will have rotated 360°. If you continue doing that and spin around twice, you will have rotated 720°. Spinning around three full times will take you through an angle of 1080°. If you keep at it and spin around several more times, you will have made yourself dizzy. To be completely mathematically accurate, when you spin counterclockwise, the size of the angle is represented by a positive number. Technically, if you are spinning clockwise, you should say that your angle measurement is negative. Normally this doesn't come up very much, but it's worth pointing out. Your teacher might even have forgotten about this convention, so if that's the case don't push it. It'll be our secret.

If a full rotation is 360°, then rotating halfway around would be half of 360°. Remember, the word "half" means multiplying by ½, so half of 360° will be ½ × 360° = 180°. A quarter turn would be ¼ × 360° = 90°.

Whenever you see angles described, you almost always will see them given in terms of their degree measurement. Problems in physics and engineering classes usually are stated in terms of the degree measurement of an angle. Angles in aviation, like the angle that the airplane makes compared to heading north, are given in degrees. The angles of runways also are based on degree measurements.

The other way that angles are measured is called radians. You may see a button on your calculator that converts angles to radians. Don't worry about it for now. You won't be seeing radians for a few years. If you plan to take calculus in a few years, however, expect to see a lot of them. Calculus only works in radians; it doesn't work in degrees.

## Using a Protractor

A protractor is a semicircular bit of clear plastic marked off with tick marks and numbers; it's used to measure angles. There are a few features of your protractor you'll need to locate in order to use it to accurately measure angles. One is a line near the bottom (the straight side) that is marked with the numbers 0 and 180. When measuring an angle with a protractor, you need to align this line with the initial side of your angle. The other essential feature of your protractor is a dot or a cross near the center of this line. This mark needs to be lined up with the vertex of the angle you're measuring.

It is absolutely essential to get your protractor lined up accurately in order to get the correct measurement. Once you have it lined up, look to see where the terminal side of the angle crosses the curved edge of the

> **Kositsky's Cautions**
>
> Don't line up the angle with the very bottom of the protractor unless that's where the 0-180 line is. Otherwise, it's really unlikely you'll get the correct angle measurement.
>

protractor. Use the numbers along the edge to find the measurement of the angle. If your angle comes out between two numbers, you can use the tick marks between the numbers to figure out what the measure of your angle is.

GOOD

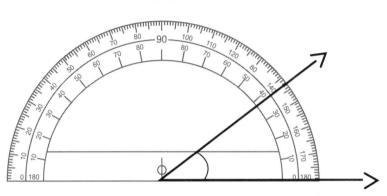

BAD

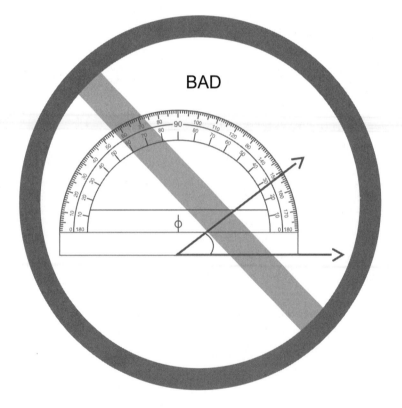

Sometimes your line segments are too small for your protractor, like in the following figure.

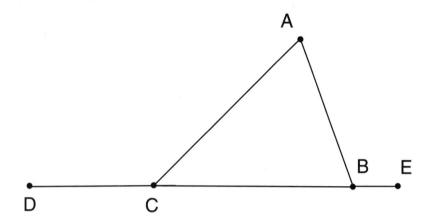

If this is the case, you need to extend the line segments using a straightedge, the mathematical term for anything that doesn't bend. After you extend it, you then can measure the angle as normal.

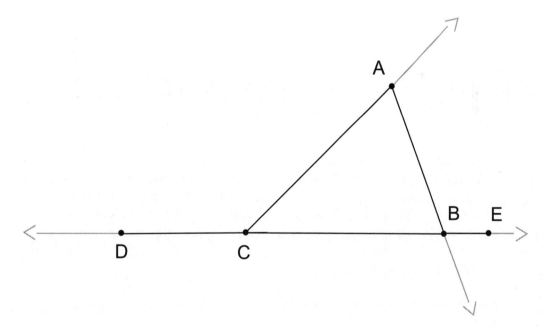

**Example:** Measure the angles ABC, BCA, CAB, ACD, and ABE in the following figure.

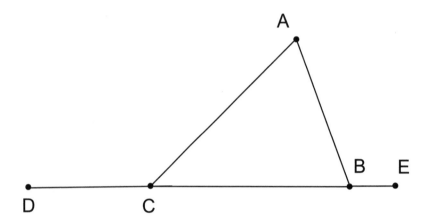

**Solution:** To use a protractor of your own:

1. Extend each of the lines on the diagram, as shown in gray.

2. Place your protractor on the figure so that the 0° and 180° tick marks both lie on the line going through A and C or both lie on the line going through B and C. In this case we chose to place the 0° and 180° tick marks on the line going through B and C.

3. Look at where the line passing through A and C intersects the tick marks. On the outer set of numbers the intersection is between 130 and 140, about four tick marks to the left of 140. That intersection means our angle might be 136°. On the inner set of numbers it is between 40 and 50, about four tick marks to the left of 40. That intersection means our angle might be 44°. Which one is it? If we were standing at C looking at B, then we turned left until we were looking at A, it would be *less* than a quarter turn. This measure of angle BCA must be less than 90°. So the angle BCA is 44°.

Using the same method, we calculate that ABC = 69°, CAB = 67°, ACD = 136°, and ABE = 111°.

## Estimating Angles

Once you are comfortable with measuring angles with a protractor, you'll start to get the intuition that you need to be able to estimate angles. One of the easiest angles to recognize is one that is about 90°. You should be able to identify them on sight. Also called right angles, they are the angles that live in the corners of squares and rectangles and many everyday objects that are constructed by people.

| Practice Makes Perfect |
| --- |
| Problem 1: Try to get your hands on a clock. What is the angle between the minute hand and the hour hand at 12:20? Pretend that the hour hand stays exactly on the 12. |

Beyond that, you need some benchmarks. If you take a square piece of paper and fold it diagonally in half (corner to corner) and then cut along your fold line, that will give you a 45° angle. Two 45° angles fit together to make a right angle. If you take one of your 45° angles and put it next to a right angle, you're up to 135°, which is the angle you'll find in the corner of a stop sign. From here, you can estimate a fair number of angles.

Beyond that, you can take your protractor and draw angles that measure 30°, 60°, 120°, and 150° and learn what they look like. Try to find objects around you that match up to these measurements. Once you've guessed the measurement of an angle, then you can check your answer by measuring with your protractor. If your two values are widely different, then either your estimate was not so good or else you measured badly. That's the beauty of having two ways to work a problem; if you get two different answers, you know you must have made a mistake somewhere.

# Types of Angles

When we were estimating angles, we began by looking at a 90° angle. A few reasons why we started with that: it's the most common angle, the most easily recognized angle, and the one that has the most talked-about mathematical properties. It's sort of a superstar of angles. But let's not forget the little people. It's time to add some more terms to our taxonomy of angles.

Remember, I told you that mathematicians are like biologists—trying to classify everything and give it names. I know that you probably regret the fact that every time we use the word "classify" we are not making things more classy nor trying to repair the bad reputation that mathematics has when it comes to style. The only class that we have is math class. In any event, there are three types of angles that you will need to know about: acute, obtuse, and right.

## It's an Acute Little Angle

One type of angle you'll come across is the acute angle. In general terms, these are angles that are less than (but not equal to) 90°. More specifically, when you measure an angle and find that its measure is between 0° and 90°, we call this an acute angle. (The restriction of it having to be greater than 0° is there just in case you inhabit a mathematical realm that involves negative angles. They exist, but not everybody has to deal with them.)

*Examples of acute angles.*

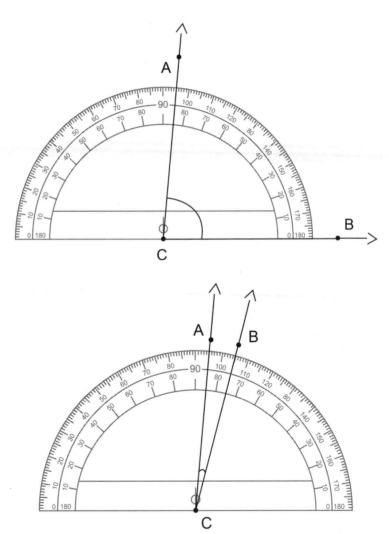

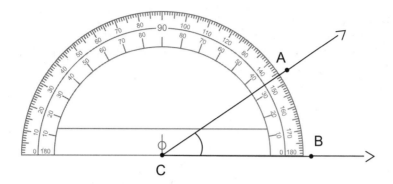

Acute angles are sort of sharp and pointy. They're the types of angles that you might get yelled at for playing with in the house, with the vague warning that someone might put an eye out. Often, when someone draws an angle, it's likely to be acute.

## Don't Be Obtuse

The next type of angle you'll encounter is one that opens wider than 90°. When you measure an angle and find that its measure is between but not equal to 90° and 180°, we call this an obtuse angle. Obtuse angles have a somewhat gentler bend than the acute angles. They're not quite so pointy or dangerous.

*Examples of obtuse angles.*

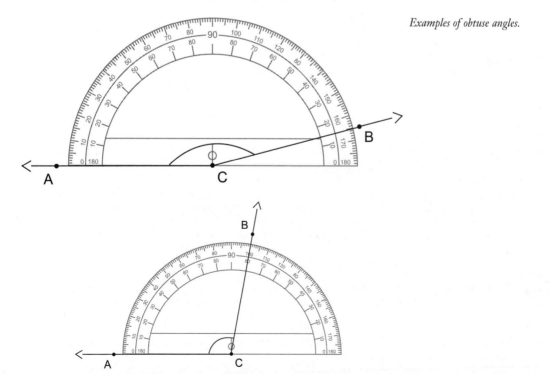

## The Right Answer

So what about our superstar, all-present, attention-grabbing angle? What happens if the measure of your angle is exactly 90°? We've saved the best for last. The 90° angle comes up so often that we have a special name for it. Any angle that measures 90° is called a right angle.

This name should signify positive connotations, good luck, and much fame and fortune. That's actually pretty much on the spot, as right angles are pretty highly valued. Unless you're building towers that lean on purpose, carpets that don't quite fit the room, or windows that don't shut completely, you're going to need to have right angles.

*Examples of right angles.*

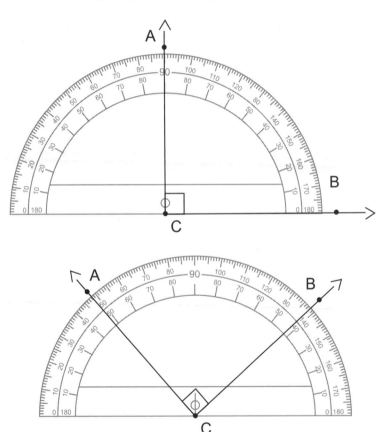

## Special Pairs of Angles

Remember how I said that the right angle was the superstar of angles? Like a rock star combined with a movie star combined with an influential politician. It's everywhere, and you can't escape it. Most things meet at right angles. When houses are built, a great deal of effort is made to make sure that the walls all meet at right angles—except in buildings with weird modern architecture where the walls meet at funny angles on purpose. It's easiest to drive when the roads meet at right angles.

The other angle that comes up a fair amount is a 180° angle—which is like having no angle whatsoever. You might be able to imagine a situation where things meet at an angle and you wish that they didn't and you wanted everything to be straight and lined up.

In order to look at these types of situations, sometimes we look at pairs of angles taken together. The two cases we examine are called complementary and supplementary angles.

## Complementary Angles

You might think that complementary angles are angles that say nice things about you. Not the case. Those would be called complimentary angles (with an i). What we're looking at are complementary angles (with an e). The mathematical idea of complementary angles is any two angles whose measurements add up to 90°. They are called complements of each other.

**Example:** You have a 30° angle. Find its complement.

**Solution:** Because complementary angles add up to 90°, you need to find out 30° plus what equals 90°? This begs to be put into an equation. Let $x$ be the measurement of the unknown angles: $30° + x = 90°$. Subtracting 30° from both sides and solving for $x$, you get $x = 60°$.

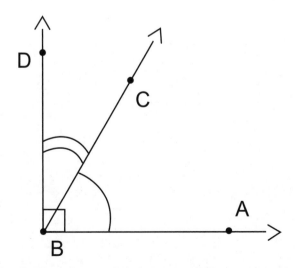

# Supplementary Angles

These are somewhat deceptively named, too. Don't think about these as angles that provide extra vitamins to your food and supplement its nutritional value. Supplementary angles are angles whose measures add up to 180°. They are sometimes called supplements of each other.

**Example:** You have a 120° angle. Find its supplement.

**Solution:** Let $x$ stand for the measure of the unknown angle. Then $120° + x = 180°$. Subtracting 120° from both sides and solving for $x$, you have $x = 60°$.

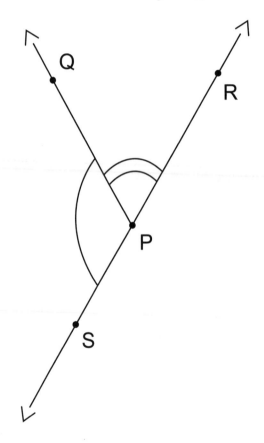

 **Timely Tips** _____

C comes before S in the alphabet and 90 comes before 180 on the number line. The sum of complementary angles is 90° and the supplementary sum 180°. C before S, 90 before 180 is a great way to remember which is which.

# Other Types of Angles

There are a few other terms that might come up when you're learning about angles. These are mostly the sorts of things that will come up in trick questions and other special cases. Still, to be fully prepared for whatever is coming at you, you probably need to know what these are.

The main thing going on with these three types of angles is whether they measure less than 180°, more than 180°, or exactly 180°. It's pretty easy to split off the case when an angle measures exactly 180°, and it's a bit more subtle to tell the difference between an angle that measures less than 180° and one that measures more than 180°. I know, at first that probably seems crazy. You are probably pretty good at telling the difference between angles that measure more than 90° from those that measure less than 90°; you can probably identify obtuse and acute angles from across the room.

The tricky part here is that you need to know where the inside of the angle is and where the outside of the angle is. This is where it becomes important to know which ray is the initial side and which ray is the terminal side of the angle. Remember, angles are always measured starting at the initial side and ending at the terminal side.

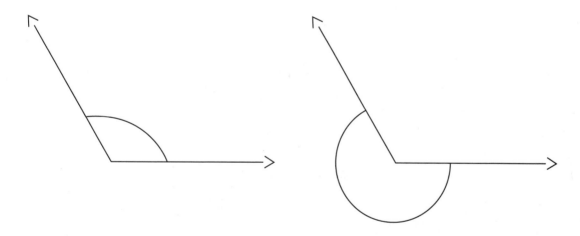

## Convex

When we talk about a shape being convex, we're talking about its angles. When all of the angles in a shape measure less than 180°, then the shape is said to be convex. Most of the normal shapes you are familiar with are convex. In fact, there'd be no need to introduce terms like convex if it weren't for the fact that there are a few weird situations that need to be named.

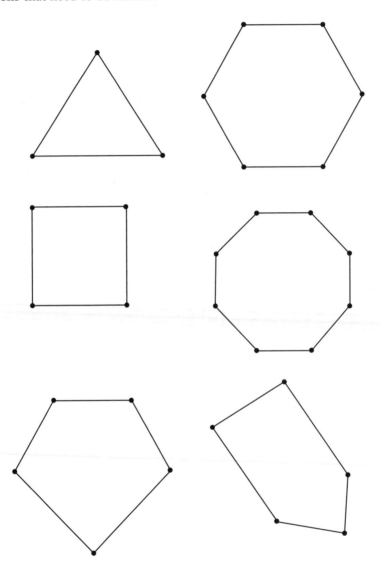

## Concave

In some sense, the opposite of convex is concave. A shape is concave if any of its angles measure more than 180°. The word "con" in Spanish means "with" and "cave" in English means, well, "cave." You can think of concave figures as those with caves in them. Any figure with a single indentation or a whole network of caverns is concave.

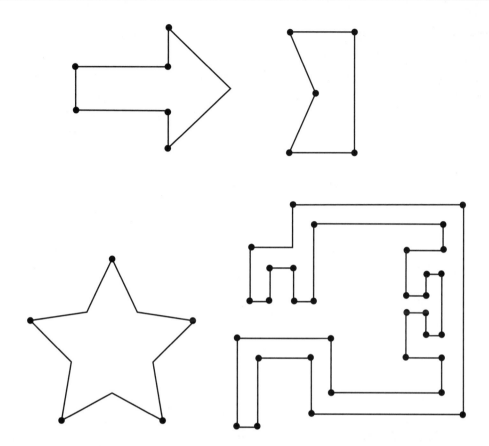

## Straight

This is the sneakiest kind of angle out there. It doesn't even look like an angle. It looks like a straight line. Maybe it will hint to you that it's an angle by having its vertex labeled with a letter. A straight angle measures 180°. Most people don't pay attention to straight angles, so these angles usually feel left out.

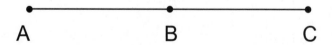

A          B          C

# Parallel and Perpendicular Lines

When lines meet, there aren't a lot of options. It's not like people who have several choices of where and how to meet: go for coffee, see a movie, or play Frisbee in the park. When lines meet, all they tend to do is to intersect in a point. And they don't always meet. Some lines want nothing to do with each other and keep their distance. It doesn't need to be a large distance, but it's there.

Our language about angles gives a less metaphorical way to talk about lines and the ways that they may or may not intersect. When you have two lines in a plane, they'll either meet or they won't. Lines that do meet will do so in an angle. When the angle is a right angle, then we say that the lines are perpendicular. No matter what angle they meet at, the lines will form two pairs of supplementary angles.

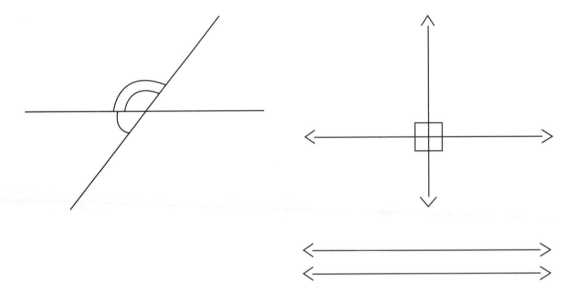

What happens if the lines don't meet? Now, when I say that they don't meet, I mean that it's not that they're having a secret rendezvous and getting together somewhere off the edge of the paper. They aren't heading toward each other gingerly, carefully, hesitantly like two people who like each other but who are afraid to admit it. No, these are lines that never intersect, that never touch, that never move closer together. These are parallel lines. They go on forever, always the same distance apart, never meeting.

## The Least You Need to Know

- You can measure angles with a protractor or estimate them visually.

- Acute angles are less than 90°, right angles equal 90°, and obtuse angles are greater than 90°.

- Complementary angles add up to 90°. Supplementary angles add up to 180°. C before S, 90 before 180.

- Parallel lines do not intersect, and perpendicular lines intersect at right angles.

# Poly(gon) Gets a Cracker

## In This Chapter

- Different types of triangles and rectangles
- Shapes with more than four sides
- Classifying shapes based on their sides and angles
- Measuring around a shape
- The Pythagorean Theorem

When people talk about "getting in shape," they probably aren't thinking about geometry. But why not? Math gives you a mental workout. Don't you think that there is a reason that the homework problems in many math books are called "exercises?"

Now that you know how to flex your angles, you're ready to build up a body of shapes. In this chapter, you'll be working out the properties of triangles, rectangles, and other flat-sided shapes.

## Classifying Shapes

Mathematicians have an insatiable need to sort and classify things. It's like the kid who takes all his toy cars and separates them by kind and then sorts them again by color and then reclassifies them again by some other trait. You should remember from Chapter 1 that mathematicians have come up with a scheme for categorizing all the types of numbers like integers, rationals, and reals. Shapes are not exempt from this need for order, and there are names for all different types of shapes based on the features that they share.

## Types of Triangles

You probably are pretty familiar with triangles. Three sides, three angles, what more is there to say? Apparently, a lot. The simplicity and elegance of the triangle make it ubiquitous in mathematics. Just like poets write odes to beautiful women, mathematicians describe the properties of triangles.

There are a few basic types of triangles that are defined by the lengths of their sides:

- Equilateral triangle. The name sort of gives it away—"equi" as in equal, "lateral" as in side. If all three sides of a triangle are the same length, then it is an equilateral triangle. It works out that if the sides are all the same length, then all three angles have the same measure and are all 60°.

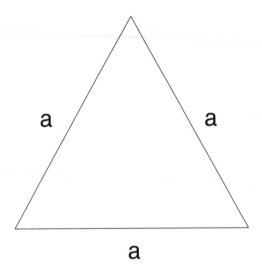

- Isosceles triangle. These triangles have at least two sides that are exactly the same length. They also have a pair of angles with exactly the same measure. I'm not exactly sure where the name isosceles comes from, but mathematicians tend to use the prefix "iso-" when things are the same.

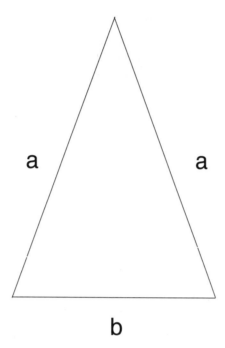

♦ Scalene triangle. You can probably guess the pattern here. In a scalene triangle, all three sides have different lengths and all three angles have different measures. I dare you to name your daughter Scalene.

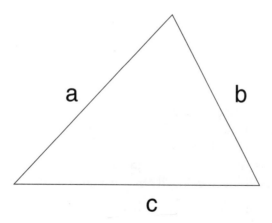

These basic classifications of triangles are based primarily on the lengths of their sides. Sure, there are also consequences in terms of the angles of the triangles, but when we're thinking about equilateral, isosceles, and scalene triangles, we're usually thinking about side lengths.

- Right triangle. There is one other important kind of triangle, the right triangle. This type of triangle is defined by having a right angle in it. If one of the angles of a triangle measures 90°, then it is a right triangle. You might wonder what we would call a triangle with two right angles, and the answer is that we don't call it anything because such a triangle is impossible. You'll see more about that in geometry class.

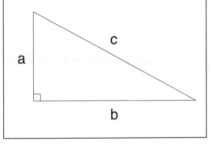

Right triangles have some extra vocabulary attached to them. The longest side of a right triangle, which is always located across from the right angle, is called the *hypotenuse*. The other two sides, the ones that are next to the right angle, are called the legs.

There isn't much to calculate about these shapes and most of the questions you might be asked about them probably strike you as somewhat philosophical. Can an equilateral triangle be an isosceles triangle? Can a right triangle be an isosceles triangle? Can a right triangle be an equilateral triangle?

The answers are yes, yes, and no.

The definition of an isosceles triangle says that it must have two sides that are the same length but it does not specify anything about the third side. If it's the same as the other two, that's fine. If it differs from the other two, that's cool too. Just as long as you have two sides that are the same, the triangle is isosceles. With this interpretation, an equilateral triangle is definitely isosceles.

It's perfectly legit for a right triangle to be an isosceles triangle. This is, in fact, a special type of triangle called the 45-45-90 right triangle and you'll see a lot of it when you study trigonometry. What can't happen, however, is an equilateral right triangle. In an equilateral triangle, all the angles measure 60°. In a right triangle, you need an angle that measures 90°. If they're all 60°, then none of them can be 90°, so an equilateral triangle can never be a right triangle.

| Practice Makes Perfect |
|---|

Problem 1: A triangle has sides of length 4, 4, and 3. What kind(s) of triangle is it?

Problem 2: What kind(s) of triangle is this?

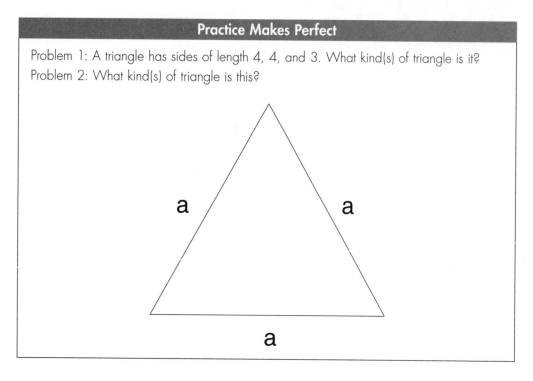

## Types of Quadrilaterals

While you have definitely heard about triangles and have thought about them before, you might not recognize the term "quadrilaterals." This is, once again, the intrusion of ancient languages into mathematical nomenclature. "Quad-" means four and, as we said earlier, lateral means side. Thus, a quadrilateral is a shape with four sides.

You're probably thinking, "Oh, like a square?" Yes, a square is a type of quadrilateral, but that's not the only one. There are many more. More confusingly, some of their names depend on where you lived; the names vary depending on whether you're in North America or if you live elsewhere.

Here is a rundown of the quadrilaterals that you will see in math class:

- ◆ Rectangle. All of the angles in a rectangle are right angles. The opposite sides of a rectangle are the same length and they are parallel.

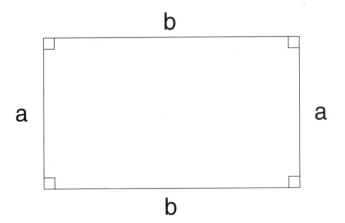

◆ Square. This is a special type of rectangle where all the sides are the same length.

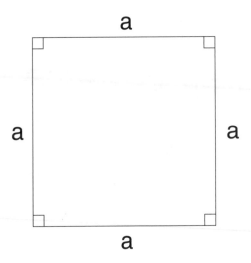

◆ Parallelogram. The opposite sides of a parallelogram are the same length and the opposite sides are parallel. A rectangle is a special type of parallelogram that has right angles for its corners.

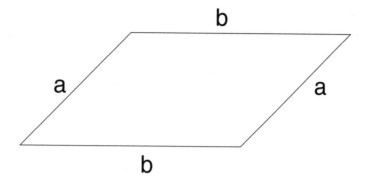

♦ Rhombus. A rhombus is a parallelogram where all the sides have the same length. A square is a special type of rhombus that has right angles at every corner.

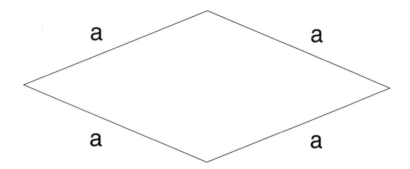

♦ Trapezoid. In North America this is a quadrilateral that has one pair of parallel sides. In British English, this shape is called a "trapezium." There is disagreement in some circles about whether a trapezoid has exactly one pair of parallel sides or at least one pair of parallel sides. If you're taking a class or course in pre-algebra, ask your teacher whether or not a parallelogram is a trapezoid in your world.

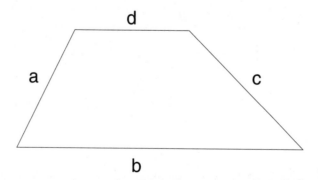

◆ Trapezium. In North America, this is a quadrilateral that doesn't have any parallel sides. In British English, this shape is called a "trapezoid." You'll notice that the words "trapezoid" and "trapezium" have opposite meanings depending on where you live. While this makes it easy to get confused, it also gives you an excuse to argue a wrong answer: just pretend that you're from somewhere else.

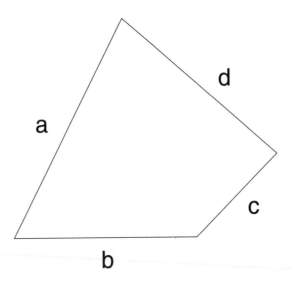

◆ Kite. This is a trapezium that has two pairs of sides that are the same length.

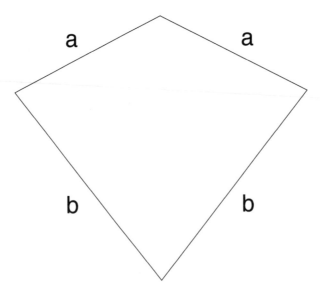

And there are other quadrilaterals that don't have common names.

The philosophical questions that you might have been asked about triangles only get worse when you move on to the realm of quadrilaterals. If a trapezoid falls in the forest and there is no one to hear it, does it make a sound?

**Example:** Is every rhombus a square?

**Solution:** No. While every square is a rhombus, it's not true the other way around. A rhombus doesn't have any restriction on the measure of its angles, and a square must have all right angles.

**Example:** Is every rhombus a parallelogram?

**Solution:** Yes. A rhombus is a special type of parallelogram.

**Example:** Is every parallelogram a rhombus?

**Solution:** No. There are no restrictions on the lengths of the sides of a parallelogram, but all of the sides of a rhombus must be the same length. While opposite sides are parallel, not all of the sides must have the same side length in a parallelogram.

> **Practice Makes Perfect**
>
> Problem 3: Is every kite a rhombus?
>
> Problem 4: Is every trapezoid a kite?

# Polygons

We can continue to look at shapes, considering five-sided, six-sided, and other many-sided shapes. Whenever you have a shape with flat sides (no curved edges) that can be drawn on a sheet of paper, we refer to it as a polygon.

Most polygons are named after their number of sides, typically with a Greek or Latin prefix for the number of sides, followed by the suffix -gon. Two obvious counter-examples to this naming rule are the triangle (not called a trigon) and the quadrilateral (not called a quadrigon). The rule picks up at five sides with the pentagon; at six sides you have the hexagon. If you're not up on your Greek and Latin prefixes for numbers, there's another way to name a polygon. If you forget that a twelve-sided polygon is called a dodecagon, you can just call it a 12-gon. Similarly, a 30-sided figure can be called a 30-gon. I suppose that you could call a quadrilateral a 4-gon, but that sort of seems like showing off. If that's your cup of tea, sip it with caution. It may be more scalding than you think to yourself or others!

**Kositsky's Cautions** _____

There are two things we don't allow polygons to do. The first is have their edges cross each other. The second is to have more than two edges meet at any one point. If they do this, they're thrown out of the polygonal club.

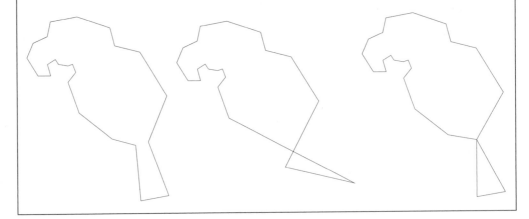

## Regular Polygons

I don't mean to sound elitist, but there are some shapes that get more respect than others. The regular polygons are the ones where all the sides have the same length and all the angles have the same measure.

You've already seen some of these shapes: the equilateral triangle is a regular polygon because its sides are all the same length and each of its angles measures 60°. The square is a regular quadrilateral because its sides are all the same length, and each of its angles measures 90°. There's no reason to stop at four sides. You can have regular pentagons, regular hexagons, and so on. No matter how many sides you specify, you can have a regular polygon with that many sides.

One of the things that mathematicians like about regular polygons is that they have a lot of symmetry. See Chapter 13 for a refresher on symmetry if you'd like one. Every regular polygon has rotational symmetry, if you rotate it around its center, as well as reflection symmetry. The regular polygons are the most symmetric flat-sided shapes that we can draw. The only shape in the plane with more symmetry is the circle. In fact, if you try to draw a regular polygon with a zillion sides, it will end up looking a lot like a circle.

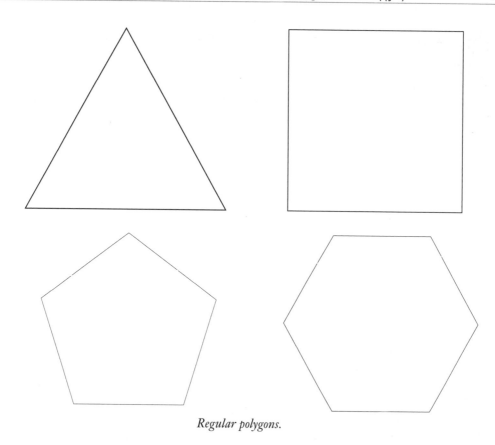

*Regular polygons.*

## Irregular Polygons

Anyone who has done any outlet shopping knows that "irregular" means that there is something wrong with a garment. What's wrong with irregular polygons? Nothing, really. It's just an unfortunate name. An irregular polygon is any polygon that's not regular. This category would include most rectangles. And, in truth, just about any polygon those of us without much artistic talent try to draw by hand will end up looking fairly irregular.

## Interior Angles

One question that comes up about polygons is, what are the measures of their angles? For some polygons, like the equilateral triangle, you've already been told the answer. Every angle in an equilateral triangle has a measure of 60°. Squares and rectangles you already know about, too. Their angles all measure 90°. What about other polygons? Do we know anything about the measure of the angles in a triangle in general?

Is there anything that can be said about the measures of the angles in a regular 16-gon?

As you might suspect, the answer is yes. By the power of mathematics, you can know the measure of all the angles in a regular polygon without even picking up your protractor. If you know enough math, you can be lazy about figuring things out. Eratosthenes was able to measure the distance around the Earth without leaving his lawn chair.

Let's take a look at how to find the size of the angles in a polygon. For an irregular polygon, you'll know what the angles add up to, but you'll need more information to know something about the individual angles.

The first thing you need to know is that the sum of the measures of the angles in a triangle adds up to 180°. You'll need to know this every now and then in the future, so it's best to commit it to memory now. Not only does it come up seemingly at random during math problems that you never even imagined being related to triangles, but it's sometimes a question on game shows and in trivia contests.

Finding the sum of the angles in a shape with four, five, or more sides can be done by remembering that the measures of the angles in a triangle add up to 180°, then splitting the shape up into triangles.

If you take any quadrilateral and sketch in a line that connects opposite corners, you will have divided the quadrilateral into two triangles. The sum of the measures of the angles in each triangle will be 180°. Since we have two triangles, the sum of the measures of the angles of the quadrilateral will be $2 \times 180° = 360°$. This means that in any quadrilateral the sum of the measures of the angles will add up to 360°. This agrees with what we already know about squares and rectangles. Squares and rectangles both have four 90° angles and $4 \times 90° = 360°$. When you get the same answer both ways, then you can be pretty sure that you got things right.

**Example:** A quadrilateral has angles that measure 70°, 80°, and 160°. What is the measure of the fourth angle?

**Solution:** Let $x$ represent the measure of the fourth angle. We know that all four angles add up to 360°, so $70 + 80 + 160 + x = 360$. Simplifying, $310 + x = 360$. Subtracting 310 from both sides of the equation, we get that $x$, and the measure of the fourth angle, is 50°.

What about the sum of the measures of a pentagon? If you connect corners in a pentagon, you first divide it into a triangle and a quadrilateral. Connecting two more corners turns the quadrilateral into two triangles, giving you three triangles total.

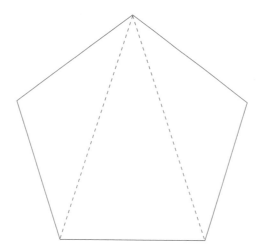

This means that a pentagon's angles have the same measure as those of three triangles, or 3 × 180° = 540°. This pattern will continue as the number of sides on your figure increases. The sum of the measure of the interior angles will be (number of sides – 2) × 180°.

**Example:** What is the sum of the measure of the angles of a 20-gon?

**Solution:** A 20-gon has 20 sides. Using the formula, (20 – 2) × 180° = 18 × 180° = 3,240°.

When you're working with a regular polygon, you can use this formula to find the measure of each angle. Remember that in a regular polygon, all of the angles have equal measure.

**Example:** What is the measure of each angle in a regular octagon? (You probably see octagons every day in the form of stop signs.)

**Solution:** This is a two-step problem. First, you need to find the sum of the measures of all the angles in the octagon and then you need to divide by 8 to find the measure of each angle. Using the formula, the sum of the measures of the angles will be (8 – 2) × 180° = 6 × 180° = 1,080°. Dividing by 8, each angle will have a measure of 135°.

> ### Practice Makes Perfect
>
> Problem 5: What is the measure of each angle in a regular 180-gon?
>
> Problem 6: The sum of a regular polygon's angles equals 720°. How many sides does this polygon have?

In case you forget any of these rules, here's a table of information for your later reference.

| Polygon Name | Number of Sides | Degrees |
|---|---|---|
| Triangle | 3 | 180 |
| Quadrilateral | 4 | 360 |
| Pentagon | 5 | 540 |
| Hexagon | 6 | 720 |
| Heptagon | 7 | 900 |
| Octagon | 8 | 1,080 |
| Nonagon | 9 | 1,260 |
| Decagon | 10 | 1,440 |

# Perimeter

You've probably watched an action movie at some point and as the danger increases and rain is pouring down, a camouflage-clad character anxiously barks into his walkie-talkie, "Secure the perimeter!" What he means is that the squad should make sure that none of the enemy are infiltrating the boundaries of their installation. We're hoping there's nobody yelling into walkie-talkies in your class, but you'll still need to be mindful of the perimeter. In this case, you'll be looking at the boundaries of a shape. For now, we'll stick to polygons. For dealing with circles and other curved shapes, you'll have to go forward to Chapter 16.

The perimeter of a shape is simply the distance around the outside of it. If you were putting a fence up around your shape, it's how much fence you would need.

## Amy's Answers

The perimeter is the distance around the outside of the shape. There's a mall in Georgia called the Perimeter Center, which is a funny name because the perimeter should be on the outside—not in the center!

It's relatively easy to calculate the perimeter of a shape. There is a two-step process and if you follow it you'll be fine. First, you find the length of each side of your shape. Then, you add up the lengths of the sides. That's all there is to it. If you follow those two simple steps, you always will be able to calculate the perimeter.

Now, sometimes there are things out there that might trick you. Sometimes the lengths of the sides are written on the diagram, which makes things easy.

Sometimes the lengths are hidden in the description of a word problem. The lengths might be given as part of a multi-step problem. You might need to use a ruler to measure the lengths of the sides. No matter what type of problem it is, though, finding the lengths of the sides is your first step. Once you know those, you can add them up.

**Example:** A hexagon has sides of lengths 2, 2, 3, 3, 4, and 5. Find the perimeter of the hexagon.

**Solution:** Since you're given the side lengths, just add them up: 2 + 2 + 3 + 3 + 4 + 5 = 19.

**Example:** What's the perimeter of a square with side length 7?

**Solution:** In this problem you need to use the fact that all sides of a square have the same length, so there are four sides of length 7. Thus, the perimeter is 7 + 7 + 7 + 7 = 28.

**Example:** You are building a fence to keep your dog restrained to an area shaped like a pentagon. Your dog insists that four of the sides of his pen must be 30 feet long. You have 160 feet of fence and you want to use it all. How long should the fifth side of your dog's enclosure be?

**Solution:** Whenever you see a fence in a word problem, you should be on the lookout for perimeter. Highly opinionated dogs don't give you much mathematical information. Let $x$ represent the length of the fifth side. 30 + 30 + 30 + 30 + $x$ = 160. Simplifying, 120 + $x$ = 160. Subtracting 120 from both sides, you get that $x$ = 40 and the fifth side of his enclosure should be 40 feet. Of course, if your dog is that smart, he could probably escape from any fence.

One thing that you need to be aware of when working on perimeter problems is that the two steps of solving the problem (find the lengths of the sides, then add them up) really were designed for your benefit. If you are looking at a rectangle that is drawn on a grid, there is a temptation to count the number of squares around the edge of the rectangle. This is wrong. You will need to find the length of each side and then add them all up.

---

### Practice Makes Perfect

Problem 7: A farmer has a 40-foot-by-320-foot field. How many feet of fence will he need to enclose it?

Problem 8: A mathematician wants to make a rhombus with a side length of 4 centimeters. What will its perimeter be?

# Pythagorean Theorem

People will tell you that you need to study math if you want to get a high-tech job. What they don't tell you is that a lot of the skilled trades require a fair amount of math, too. It probably takes more math to be an electrician than to get an MBA. One of the trades that uses the most math is construction. All sorts of things relating to measurement and shapes come up when something is being built.

Another thing that people don't tell you is that there aren't a lot of nice applications of math at the mid-level of difficulty. You can see a lot of uses for fairly basic math, like figuring out how much your groceries are going to cost or calculating your points in a game. And there are a lot of technical uses for the type of math that is taught to math majors in college. But there aren't a lot of nice examples in between. The Pythagorean Theorem, however, has a nice application to construction. If you watch any of those home renovation shows on cable TV, you'll probably see the Pythagorean Theorem in action.

So what is this theorem about? In a nutshell, it tells us the relationship between the lengths of the sides of a right triangle. If you label the legs of a right triangle with the names $a$ and $b$ and the hypotenuse $c$, then the lengths of the sides are related by the equation $a^2 + b^2 = c^2$. This means that if you're told the lengths of two of the sides of a right triangle, you can always calculate the length of the third side.

### Kositsky's Cautions

Be careful because this theorem only applies to right triangles. If your triangle doesn't have a right angle, you can't use this theorem to calculate the lengths of the sides. In that case, you'll have to appeal to trigonometry and use the Law of Cosines.

It's not obvious how a theorem about the lengths of the sides of a right triangle has anything to do with building a deck, but that comes from a remarkable fact about this theorem. If you're given the lengths of three sides of a triangle and if they satisfy the equation $a^2 + b^2 = c^2$, then your triangle is a right triangle. This is the property that's used in construction. With the possible exception of experimental modern architecture, builders want the corners of houses to be right angles. When they're starting to place the walls, they take measurements along the walls and set up a triangle, which they call the carpenter's triangle. If the triangle's sides satisfy the Pythagorean Theorem, then they know that the triangle is a right triangle and the walls meet at a right angle. If the numbers don't work out, then they know that they need to start over. This is something that's good to find out before you've built the entire wall.

**Example:** A right triangle has legs of lengths 5 and 12. Find the length of the hypotenuse.

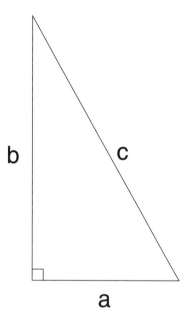

**Solution:** In the notation we've been using, the hypotenuse would be the length called $c$, while the legs would be $a$ and $b$. Setting up the equation, we get $5^2 + 12^2 = c^2$. Squaring and adding, this becomes $169 = c^2$. You can solve this either by taking the square root of both sides or by remembering that $13^2 = 169$. In either event, you'll find that $c = 13$.

**Example:** A triangle has a leg of length 8 and a hypotenuse of length 17. Find the length of the third side.

**Solution:** The hypotenuse is always the longest side of the triangle, so it's the one that we always call $c$. It doesn't matter if 8 is $a$ or $b$. You can choose either one. Let $b = 8$ and let $a$ be the unknown third side.

Setting up the equation, you get $a^2 + 8^2 = 17^2$. This becomes $a^2 + 64 = 289$.

To solve this equation, subtract 64 from each side, giving you $a^2 = 225$. This means that $a = 15$.

**Example:** A triangle has sides of lengths 10, 24, and 27. Is it a right triangle?

**Amy's Answers**

If you're interested in seeing several proofs of the Pythagorean Theorem, take a look at www.cut-the-knot. org/pythagoras/index.shtml or search the web for "Pythagorean Theorem proofs."

**Practice Makes Perfect**

Problem 9: A triangle has a hypotenuse of 15 cm and one of its legs is 9 cm. What's the length of the other leg?

**Solution:** You can check by plugging in to the Pythagorean Theorem: does $10^2 + 24^2$ equal $27^2$? $10^2 + 24^2 = 676$, but $27^2 = 729$. Since the numbers are not the same, the triangle in question is not a right triangle.

Other than calculating with the Pythagorean Theorem and using it to build your dream home, your math class may also have you prove this theorem. Unfortunately, we probably won't be able to help you with that due, in part, to the popularity and longevity of this theorem. The Pythagorean Theorem has been around for thousands of years and there are literally hundreds of proofs of it, including one done by long-ago president James Garfield. (Can you imagine any of our recent presidents spending time inventing a new proof of a mathematical theorem?) No matter which one of the proofs we would choose to describe, there are hundreds of others that your teacher might prefer. On the bright side, we will point out that this means that there shouldn't be one right answer to a question about the proof.

## The Least You Need to Know

- To name a triangle, find out how many of its sides are the same length.

- A quadrilateral is classified based on whether its opposite sides are parallel, how many of its sides are the same length, and whether it has right angles.

- The sum of angles of an $n$-gon is $(n - 2) \times 180°$.

- If you know the lengths of two sides of a right triangle, you can find the length of the third side by using the Pythagorean Theorem.

# Chapter 16

# Area, Volume, and Surface Area

## In This Chapter

- ◆ Areas of quadrilaterals and triangles
- ◆ Areas of circles
- ◆ Volumes of solids

Let's say that you were a master hunter and you had become an expert at hunting wildebeests. You've decided that you want to begin decorating your home with decorative objects made out of wildebeest hides. Maybe you want to have a wildebeest rug to cover the floor in your living room. The problem is, when you're setting your annual goals for the number of wildebeests you plan to get this year, how many will you need?

In order to figure this out, you need to know about area. You need to know how large your living room floor is as well as how much area is covered by a wildebeest hide. You'll also need to decide whether you want to have wall-to-wall wildebeest carpeting or just an area rug.

In this chapter you learn how to calculate how much space a shape takes up. Not only can you use this for interior decorating, but these ideas come up in construction, farming, and many other fields (no pun intended).

# Calculating Area

In some sense, area is how much stuff is inside a shape. In one of the oldest applications of this idea, the shapes were farms and the farmers were taxed based on the size of their land. The government wanted to know how big its citizens' farms were in order to tax them accurately, and the farmers wanted to know how big their farms were so that they could make sure that they weren't being overtaxed. Neither side really trusted the other, so they had to turn to mathematics to sort things out.

While distance might be measured in inches, meters, miles, or other such units, area is measured in square inches, square meters, square miles, or the like. A square inch, written $in^2$, is the size of a square that is one inch across. The same is true for the other measurements; a square meter (furlong, rod, stadia) is the size of a square that is one meter (furlong, rod, stadium) across. You can measure something in square units even if it isn't a square. If you imagine cutting up the square inch into small pieces, you can fit it into your shape and figure out how big your shape is. But this is math class, not arts and crafts, so put away those scissors. It's much harder to read a book that is cut up into tiny pieces!

For most of the shapes that we'll be dealing with, you'll be finding the area with formulas. Much safer than using objects banned on airplanes. No one has ever warned you not to run with math formulas or to put them in your checked luggage, have they?

## Rectangles

There are a few ways that you might find the area of a rectangle. The very simplest of these (and the one that you usually start with) is when the rectangle's sides are all whole numbers and the rectangle is filled in with a grid of squares. If the sides of the rectangle are measured in meters, then those squares are square meters; if its sides are measured in miles, then those squares are square miles. If your drawing is already presented to you in this nice, easy form, then all you need to do is to count how many squares are inside the rectangle.

**Example:** Find the area of the following rectangle by counting the interior squares.

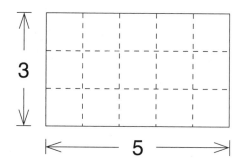

**Solution:** You can count the squares however you'd like as long as you get all of them exactly once. I usually start from the upper left and move as if I were reading. You should come up with 15 in this case.

More likely than not, you've done this sort of thing before. It might have been a very long time ago. Think back to when you were first learning about multiplication. Odds are that you saw a diagram similar to this back then. In order to learn what it meant to multiply $4 \times 6$, you may have been shown a rectangular array that was six units across and four units tall, and you found the product $4 \times 6$ by counting the number of boxes in the grid.

This is no coincidence, as the concepts of area and multiplication have some fairly close connections, and it tells us what the formula is for finding the area of a rectangle. If the base of a rectangle has length $l$ and the height of the rectangle has width $w$, then the area of the rectangle is given by the formula $A = l \times w$.

The formula has several advantages over the box-counting method. For one, you don't have to draw all the boxes. Drawing them all could take a really long time if you have a big rectangle. Additionally, the formula works nicely if the lengths of the sides are not whole numbers. Figuring out what to do with fractional boxes could definitely be a pain.

**Example:** A rectangle has sides of length 6.5 feet and 8 feet. What is its area?

**Solution:** $6.5 \times 8 = 52$ square feet.

There's one case that is particularly easy to deal with. That's when your rectangle is a square. In that case, to find the area, you take the length of the side and square it. If the side has length $s$, then $A = s^2$.

> **Practice Makes Perfect**
>
> Problem 1: What is the area of a rectangle with side lengths 0.5 and 249?

## Parallelograms

Remember that a parallelogram is little more than a smooshed rectangle. If you made a rectangle out of coffee stirrers and pipe cleaners (*What's with all the arts and crafts!?* you might ask. Actually, the authors never really left pre-school.), it would pretty easily droop into a parallelogram.

It's this droopy, smooshy nature of the parallelogram that makes it a bit different from the rectangle. The first difference is that you can't fill it with squares. Because the sides of a parallelogram stick out at an angle, if you tried to fill it with squares, they'd either stick out or else not quite fill up the shape. The other difference is that knowing only the lengths of the sides isn't going to help you.

In order to find the area of a parallelogram, you need two numbers. One of them is the length of the base of the parallelogram. The base is typically the side that is drawn horizontally on your paper, but sometimes you will need to rotate your paper (or your head) to envision the base being horizontal.

## def•i•ni•tion

Two lines are **perpendicular** if they meet at a right angle.

The other essential bit of information that you need for finding the area of a parallelogram is its height. The height is not one of the sides. Instead, the height is the distance measured from the base to the side across from it. This has to be measured in a straight direct line, *perpendicular* to the base. Perpendicular means "at a right angle to" and is one of those technical terms mathematicians use. Basically, if two lines look like they could be the corner of a square or rectangle, they're perpendicular.

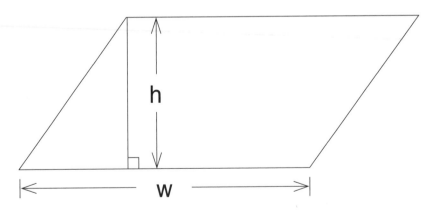

Once you have these two pieces of information, finding the area of the parallelogram is easy. You just multiply the base by the height: $A = b \times h$.

**Example:** Find the area of the following parallelogram.

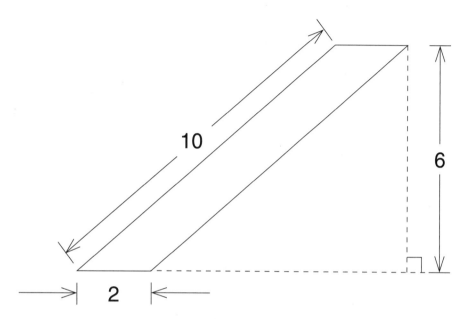

**Solution:** We need to identify a side of the parallelogram that we'll consider the base. Our two choices are the side of length 10 and the side of length 2, and it would be best to pick one that has information about the height as well. We measure the height as the perpendicular distance from our base to the parallel side. Because the distance measured as "6" in the figure fits the definition of height for the side of length 2, it makes sense to use the side of length 2 as the base. This means $b = 2$ and $h = 6$, so A = $(2) \times (6) = 12$.

It sort of makes sense that the area, that is the amount of stuff inside a parallelogram, isn't just based on the lengths of its sides. A long and skinny parallelogram can have sides the same lengths as one that is almost a rectangle. Yet the skinny one is going to have a lot less area. We see this in the previous example. A rectangle with side lengths 2 and 10 has an area of 20, whereas the particular parallelogram we looked at only has an area of 12!

---

**Practice Makes Perfect**

Problem 2: What is the area of the parallelogram below?

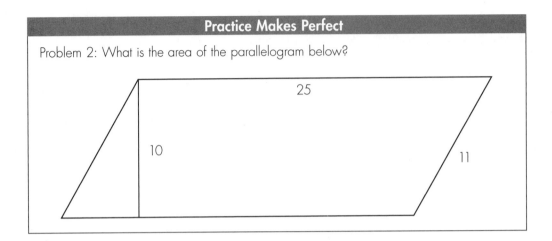

## Trapezoids

More complicated than either rectangles or parallelograms is the trapezoid. The entire trick to dealing with a trapezoid is figuring out which sides you're supposed to be working with and which sides are just there to deceive you.

A trapezoid is a four-sided figure that has one pair of sides that are parallel and the other two sides stick out whatever-which-way. The parallel sides are called the bases, and you'll use them in calculating the area of the trapezoid. One base is going to be called base-one and abbreviated $b_1$ and the other is called base-two and abbreviated $b_2$. The distance between the bases is called the height and abbreviated $h$.

Once you know these three pieces of information, you can calculate the area of the trapezoid. The formula can be written in a few different ways. All of them always will give you the same answer, and they all are mathematically equivalent. They're just phrased somewhat differently; different people prefer different versions. This is the standard version you will probably see most often:

$$A = \left( \frac{b_1 + b_2}{2} \right) h .$$

**Example:** Find the area of the following trapezoid.

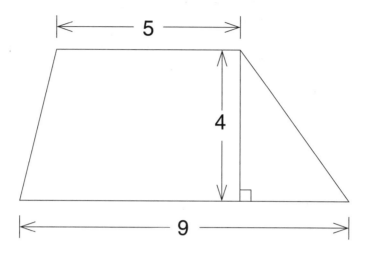

**Solution:** First, we have to decide which opposite pair of sides are parallel. The top and the bottom seem to be aligned perfectly with one another, whereas the left and right sides look like they will crash into each other if they continue their way up the page. Let's stick with the safer choice for now. Even if you're a rebel, you have the cause of math to live for!

The lengths of the bases are 5 and 9. It doesn't matter which is $b_1$ and which is $b_2$. Let $b_1 = 5$ and $b_2 = 9$. The other piece of information that you need is the height; fortunately, it's nicely labeled for you in the problem, $h = 4$.

Putting these into the formula for the area of a trapezoid, we get $\left(\dfrac{5+9}{2}\right)4$, which equals 28.

---

### Practice Makes Perfect

Problem 3: What is the area of the following trapezoid?

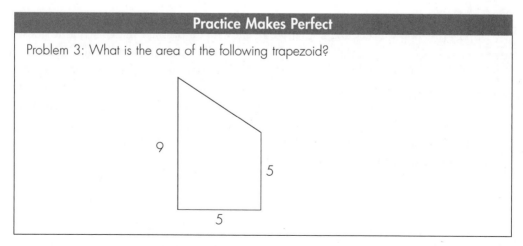

## Triangles

Up until this point, all of the examples have been with quadrilaterals. The odd man out in these area calculations is the triangle. Once again, the quantities called "base" and "height" will be put into play. Any side of the triangle can be the base, and it's typically drawn so that the base is horizontal and at the bottom of the triangle (which makes sense for something called a base). The height of the triangle is measured by drawing in the altitude. The altitude of a triangle is a line segment that connects the base of the triangle with the corner directly opposite from it; the altitude needs to be perpendicular to the base. The length of the altitude is called the height of the triangle. Just like before, we're going to have $b$ stand for the base and $b$ stand for the height.

The formula for the area of a triangle is $A = \frac{1}{2}bh$. You might be wondering where the one-half comes from. This is because a triangle is half of a parallelogram. If you draw a line diagonally corner-to-corner in a parallelogram, you will get two triangles. Because the area of the entire parallelogram is $bh$, the area of the triangle that makes up half of the parallelogram will be half of $bh$.

**Practice Makes Perfect**

Problem 4: If the hypotenuse of a right triangle is 13 m and one leg is 12 m, what is the triangle's area and the length of the other leg?

# Circles

Up to this point, every shape that we've shown you has had flat or straight sides. It's been the sort of thing that you could measure directly with a ruler, if you'd wanted to, and that you could imagine filled up with squares (or pieces of squares cut up with scissors). There's been a glaring omission here. Circles. Of all of the shapes with curved sides, circles are the ones that show up the most in math class. While you might not really care to calculate how much data can fit on a CD-ROM or how large a field can be watered with a circular irrigation system, you will be working with circles and finding their measurements throughout your mathematical career.

While the area of a circle is still called its area, there is a special name for the perimeter of a circle. The distance around a circle is known as the circumference, and it is typically denoted by $C$. Other key measurements of a circle are its diameter, $d$, and its radius, $r$. The diameter is the distance across the circle at the widest part of the circle and passing through the center. The radius is a line segment that connects the center of the circle with the edge of the circle. The length of the radius is always one half the length of the diameter.

## Easy as Pi

Before getting too far with circles, we need to introduce some more Greek vocabulary. In this case, it is the letter $\pi$, which is the lower-case letter pi from the Greek alphabet. The letter $\pi$ stands for a number whose decimal representation begins 3.14159 ... and goes on forever without repeating.

You might be wondering what's so special about this number and what it has to do with circles. There are a lot of things that are special about it. The first one is that it is an irrational number. This means that it is impossible to write it as a fraction where both the numerator and the denominator are integers. It also means that the decimal expansion of $\pi$ will keep going on forever without repeating or having any fixed pattern. This means that if you use a decimal version of $\pi$, it isn't the exact number but just an approximation. Even if you put in all the digits that will fit on your calculator screen, it is still just an approximation. Even if you use a million digits of $\pi$ in your calculation, you will still have just an approximation and not the exact answer. The more digits you use, the better your approximation will be, but it will still just be approximate. Unless you're told otherwise, you can use 3.14 as an approximation for $\pi$—just remember that it's only an approximation.

**Timely Tips**

English-speakers tend to pronounce $\pi$ as "pie" while people who grew up speaking other languages tend to pronounce it as P. The letter $\pi$ has been in use for about 300 years to represent the ratio of the circumference of a circle to its diameter.

The relationship between $\pi$ and circles is as follows: If you take any circle with circumference $C$ and diameter $d$, the ratio $\dfrac{C}{d} = \pi$.

# Getting Around

There are three equations that relate the radius $r$, the diameter $d$, and the circumference $C$ of a circle. They are:

- $d = 2r$
- $C = \pi d$
- $C = 2\pi r$

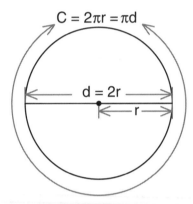

With these three equations, you can answer just about any question about distances in a circle.

**Example:** A circle has a diameter of 4 inches. What is its radius?

**Solution:** Because $d = 2r$, we plug in 4 for $d$, giving us $4 = 2r$. Divide both sides of the equation by 2 to solve for $r$, and you have $r = 2$ inches.

**Example:** A circle has a diameter of 4 inches. What is its circumference?

**Solution:** Because $C = \pi d$, plug in 4 for $d$ to get $C = \pi(4)$. The exact answer to this question is that the circumference is $4\pi$ inches. To get a decimal approximation calculate $4 \times 3.14 = 12.56$ inches.

**Example:** A circle has a circumference of 16 inches. What is its radius?

**Solution:** Here you use the equation $C = 2\pi r$. Plug in 16 for $C$ to get $16 = 2\pi r$. Now you just need to remember that 2 is a plain old number and $\pi$ is also just a number. You can divide both sides of this equation by $2\pi$ to isolate the variable $r$. When you divide both sides of the equation by $2\pi$, you get that $r = \dfrac{16}{(2\pi)}$. You can simplify that to $r = \dfrac{8}{\pi}$ inches and

---

**Practice Makes Perfect**

Problem 5: What is the radius of a circle with a circumference of $\pi$ meters?

Problem 6: Your bike's wheel has a diameter of 26 inches. How far do you travel every time your wheel rotates once?

still have an exact answer. To get a decimal approximation, you can plug in 3.14 for $\pi$, calculating the radius to be about 2.55 inches long.

## Calculating Area

The area of a circle is $\pi r^2$. Most people think that pie is round, but all mathematicians know that the formula for the area of a circle tells us that *pie are squared!* Yes, that is the sort of corny joke that you have come to expect from a math book. I hope that its embarrassing lameness will sear the formula for the area of a circle into your brain, never to be forgotten. Pie are squared! Ha, ha, ha … ha, ha … ha …

**Timely Tips**

If you have a circle and want to make a rectangle with the same area, one way to do it is to make the length of the rectangle $\pi r$ (half the circumference of the circle) and the width of the rectangle $r$.

**Example:** Find the area of a circle with radius 7 cm.

**Solution:** Using the formula $A = \pi r^2$, calculate $\pi(7)^2 = 49\pi$ cm². If you want an approximation, you can plug in 3.14 for $\pi$ and get that the area is approximately 153.86 cm².

# Shapes in Space

We aren't just limited to the shapes that are stuck down to sheets of paper. We can calculate all sorts of things about the shapes around us, things that exist in the real world as well as outside of the realm of the page and the blackboard.

These shapes are collectively referred to as solids, and you may be asked to calculate their volumes or their surface areas. The volume of a shape tells you how much is inside it.

Volume is measured in cubic feet, cubic inches, cubic meters, and so on. Just like we represented square feet by ft², we will use the symbol ft³ for cubic feet.

The other thing we measure for shapes in space is their surface area. Imagine that you were going to cover the entire outside of a shape

**Practice Makes Perfect**

Problem 7: What is the area of a circle with a circumference of 14 inches?

Problem 8: What is the shaded area of the following figure?

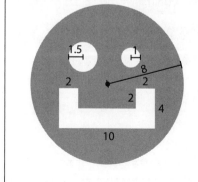

with paper. Surface area tells you the combined area of all that paper. Surface area is used when calculating how much paint you need in order to paint something or the number of shingles you need when building a roof.

# Cubes

In case your evil (or good) teachers left this out of your math education, let's start out with a definition of a cube. A cube is simply the three-dimensional version of a square. Just like a square, all the sides of a cube are exactly the same, and all the corners are made of right angles. If you were going to build a cube, you could start with six identical squares and fit them together. You see cubes in real life in dice, in perfectly square boxes, and in other such shapes. Unfortunately, ice cubes are not mathematical cubes, though; they go much better with drinks than doing math.

If you have a cube where every side is 1 foot long, then the volume of the cube is 1 ft$^3$. This is a fairly standard measure of volume, and all sorts of things—even refrigerators and microwaves—have their volumes measured in cubic feet. When finding the volumes of other shapes, you can imagine them being filled with a bunch of little cubes.

Let $s$ stand for the length of a side of a cube. The volume of the cube is given by $V = s^3$. Just like the area of a square was side length squared, the volume of a cube is side length cubed.

To find the surface area of a cube, you need to add together the areas of all six sides of the cube. Fortunately, this isn't too hard because all six sides are identical squares. If the side length of each of these squares is represented by $s$, the area of each of these squares is $s^2$. Since there are six squares, the total area of all of them together is $6s^2$, so the formula for the surface area of a cube with side length $s$ is $S.A. = 6s^2$.

**Example:** What is the volume of a cube with side length 5 meters?

**Solution:** The volume of the cube is $V = 5^3 = 125$ cubic meters (which also can be written as 125 m$^3$).

**Example:** What is the surface area of a cube with side length 5 meters?

**Solution:** The surface area of the cube is $S.A. = 6(5)^2$ = 150 square meters (which can also be written as 150 m$^2$). The units on surface area always will be square units because it is measuring an area.

> ### Practice Makes Perfect
>
> Problem 9: You have a giant plush cube-shaped chair that needs to be reanimated after an accident claimed most of its insides. If the case has a side length of 2 feet, how much foam stuffing should you order to refill the defunct chair?

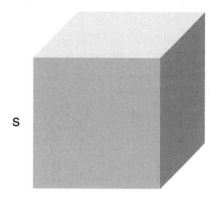

## Prisms

A prism is a three-dimensional shape made by making a box where one set of opposite sides is a polygon and the other sides are all rectangles. Most of the prisms that you will be working with will be triangular prisms and rectangular prisms, but you can make one out of trapezoids, dodecagons, or any flat-sided shape that you want to work with. This flat-sided shape is called the base of the prism. The height of a prism is the distance between the base and the side directly opposite from it. Be careful, as a prism might not be drawn with the base on the bottom!

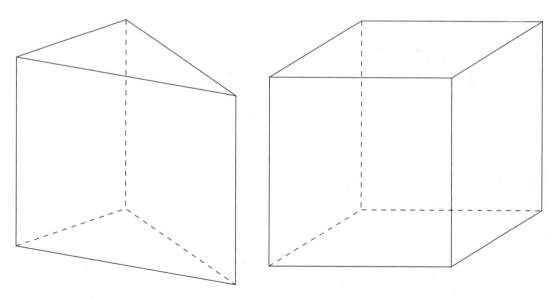

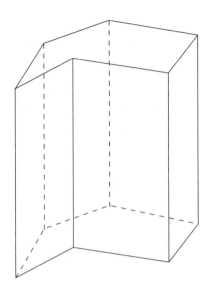

To find the volume of a prism, you multiply the area of its base times the height. As a formula this is $V = A_{\text{base}}h$.

**Example:** You have a rectangular prism. The base is a rectangle that measures 5 inches by 3 inches, and the height of the prism is 7 inches. Find the volume of the prism.

**Solution:** The area of the base is $5 \times 3 = 15$ square inches. So the volume of the prism is 15 square inches $\times$ 7 inches $= 105$ in$^3$.

What this example shows is that if you have a rectangular prism, you can find the volume fairly easily by using the formula $V = lwh$, where $l$ is the length, $w$ is the width, and $h$ is the height. The really nice thing about rectangular prisms is that it doesn't matter which is $l$, which is $w$, and which is $h$.

**Example:** You have a triangular prism. The base is a triangle with base 3 cm and height 4 cm. The height of the prism is 8 cm. Find the volume of the prism.

**Solution:** The area of the base is $(0.5)(3)(4) = 6$ cm$^2$. Multiply that by the height of the prism, 6 cm$^2$ $\times$ 8 cm $= 48$ cm$^3$.

Finding the surface area of a prism is a bit trickier. It's not that bad if you have a rectangular prism, but in other cases, it depends on a lot of stuff; there isn't an easy formula for it most of the time. You probably won't encounter any of these unless you end up in a fairly serious geometry class. By that point, you will be the master of the shapes, and the lack of an easy formula won't be a problem.

When you have a rectangular prism with dimensions $l$, $w$, and $h$, the surface area is given by S.A. = $2lw + 2lh + 2wh$.

**Example:** Find the surface area of a prism that measures 4 inches by 5 inches by 6 inches.

**Solution:** The surface area is S.A. = 2(4)(5) + 2(4)(6) + 2(5)(6) = 40 + 48 + 60 = 148 square inches.

Problem 10: What's the surface area of a rectangular prism with a volume of 30 cubic centimeters, a length of 2 centimeters, and a height of 3 centimeters?

# Pyramids

Now's your chance to get involved with some pyramid power. The ancient Egyptians had their great pyramids; in modern times we need to avoid being ripped off in a pyramid scheme. And now you will be able to find the volume of a pyramid. This might come in handy if you are time-warped back to ancient Egypt and end up being put to work on a crew building a tomb for the pharaoh. It would be embarrassing if they designed a pyramid that didn't have enough space in it for all of the wealth he needed to bring with him to the afterlife. But armed with the knowledge of how to find the volume of a pyramid, you can make sure that there is enough room for all of the gold, grain, and whatever else needs to be packed for the journey.

Pyramids, like prisms, come in different flavors. A pyramid is made up of a base, which is a polygon, and sides, which are triangles that meet at a point at the top of the pyramid. A triangular pyramid has a triangle for the base; a square pyramid has a square for the base. You can have any type of pyramid that you want, just as long as the base is a polygon. Once you have your base, just slap some triangles around the sides, make sure that their tips all meet at the top, and there you have your pyramid.

The height of a pyramid, denoted by $h$, is the distance from the base straight up to the tip. Let $A_{base}$ stand for the area of the base. We calculate the volume of a pyramid as $V = \frac{1}{3}A_{base}h$.

**Example:** The pharaoh has commissioned a pyramid to be built in his honor. He wants it to be a square pyramid with a base 230 meters on a side and to be 150 meters tall. What is the volume of the pyramid?

**Solution:** The base is a square with side length 230 meters, so the area of the base is $(230)^2 = 52{,}900$ m². Plugging this in to the volume formula, $V = (1/3)(52{,}900)(150)$ = 2,645,000 m³. That's quite a large pyramid. The pharaoh had better hope that his public works department is up to the task.

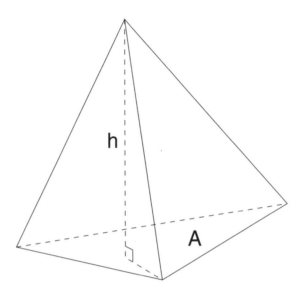

**Example:** You are making a candle shaped like a triangular pyramid. The triangle at the bottom of the pyramid has a base of 3 inches and a height of 4 inches. The height of the pyramid is 4 inches. How many cubic inches of wax will you need to make your candle?

**Solution:** The triangular base has area $(1/2)(3)(4) = 6$ in$^2$. Plugging this in to the formula for the volume of a pyramid, $V = (1/3)(6)(4) = 8$ in$^3$.

---

### Practice Makes Perfect

Problem 11: You want to build a pyramid in your own honor—in your backyard. It should be 8 feet high so all your neighbors can see it and you want it to slope gently enough so that you can climb to the top easily. This means you'll have to have the base be a 24-by-24-foot square. How much concrete will you have to buy if you want to have your monument be one solid block?

## Cylinders

When we talk about cylinders, we think about aluminum cans and barrels. These are sort of like prisms, except their bases are circles. However, because we're dealing with one case (bases are circles), they're a lot easier to deal with because there are straightforward formulas that will always work to calculate their volumes and surface areas.

When calculating things about the cylinder, we need to know the radius $r$ of the circular base as well as the height $h$ of the cylinder.

The volume of a cylinder is given by the formula $V = \pi r^2 h$, and its surface area is given by S.A. $= 2\pi rh + 2\pi r^2$.

**Example:** You have a tin can that is shaped like a cylinder. The radius of the base is 2 inches, and the height of the can is 4 inches. Find the surface area of the can as well as its volume. Give your answer as a decimal approximation by using 3.14 as the value for $\pi$.

**Solution:** Using the formula for surface area, S.A. $= 2\pi(2)(4) + 2\pi(2)^2$. Multiplying this out, you get 75.36 in². The volume of the cylinder is $V = \pi(2)^2(4)$. Multiplying this out, you get 50.24 in³.

> **Practice Makes Perfect**
>
> Problem 12: You have a cylindrical tank 5 feet long that has a base with a radius of 2 feet. What is its volume? If water weighs 62.5 pounds per cubic foot, how much would the tank weigh if you fill it with water?

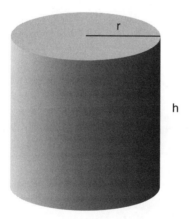

# Cones

Finally, we get to cones. Is it a waffle cone or a sugar cone? Single scoop or double scoop? Regardless of what other people might tell you, these are the important questions about cones. Mathematically, a cone is not exactly the same as an ice-cream cone. But it's close enough that your mental image of a cone can be an ice-cream cone. Just as long as you imagine it upside-down and without any ice cream in it. Or if there was ice cream in it, there isn't anymore because you turned it upside-down. Admittedly, that's a fairly sad excuse for a cone. But after learning that ice cubes are not really cubes, you shouldn't be expecting your math problems to involve real foods that come from the freezer.

To find the volume of a cone, you need to know the radius $r$ of its circular base as well as the distance from the base to the tip of the cone, which is the height $h$. Once you know these two pieces of information, you can use the formula for the volume of a cone $V = (\frac{1}{3})\pi r^2 h$.

**Example:** A cone has a base with radius 2 cm and a height of 9 cm. What is the volume of the cone?

**Solution:** Using the volume formula, plug in $V = (\frac{1}{3})\pi(2)^2(9) = 12\pi$ cm$^3$. That's the exact answer. To find the decimal approximation, you can substitute in 3.14 for $\pi$, showing that the volume is approximately 37.68 cm$^3$.

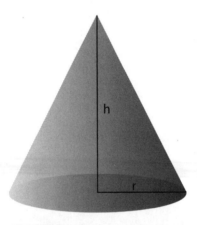

| Practice Makes Perfect |
| --- |

Problem 13: You have a square pyramid and a cone with the same height and volume. What is the ratio of the surface area of their bases?

## The Least You Need to Know

- The area of a rectangle is $lw$, the area of a parallelogram is $bh$, and the area of a triangle is $\frac{1}{2}bh$.

- The volume of a rectangular prism is $lwh$.

- For a circle, $C = \pi d$ and $A = \pi r^2$.

- The volume of a prism or a cylinder is the area of the base times the height.

- The volume of a pyramid or a cone is one third the area of the base times the height.

# Chapter 17

# Proportions

## In This Chapter

- ◆ Congruence and similarity
- ◆ Scale factors: Length
- ◆ Scale factors: Area
- ◆ Figuring out maps

You've seen them in this book and in others. Diagrams of rectangles labeled with sides with lengths like 5 meters and 7 meters. You know that the rectangle can't possibly be 5 meters by 7 meters and still fit in the book. It's a huge rectangle! It's bigger than my living room! It can't possibly fit on the page. But you recognize that the picture in the book is just a scaled down representation of a rectangle, not the actual rectangle itself.

It's the same thing when you see toy cars or model airplanes. They've taken the large object and made a smaller version of it that has all the same visual features of the original. The same idea applies to maps. You can have a map of the entire state of Rhode Island in your glove compartment. Even though Rhode Island is the smallest state by area, the actual state is still pretty big. If you tried to fold up an actual-size version of Rhode Island, it probably wouldn't work very well.

For all of these features, be they rectangles, model cars, or maps, sometimes we need to know more about the relationships between the actual object and our smaller model of it. In this chapter you'll learn about how to use the small version of the shape to calculate what's going on in the original (and vice versa).

# When Shapes Are the Same

Who knew that the idea of "same" needed to be explained in detail? You would think that things either are the same or else they aren't. By this point you definitely should expect a mathematical definition to be extremely precise and for mathematicians to nitpick all the details of what it means for two shapes to be the same. Are things the same if they're arranged differently on the page? Are they the same if they look pretty much the same but are different sizes? What about different sized squares or different sized circles? They're all the same (squares, circles), but they're still different from each other in some way.

Mathematicians use the word *congruent* to describe two shapes that are the same. For the shapes to be congruent they have to be both exactly the same shape and exactly the same size. However, it's okay if they are mirror images of each other, and it's okay if they're moved around on the page so that they're pointing in different directions. In order for two shapes to be congruent, if you cut them both out of paper, it should be possible to line them up exactly even if you need to flip them over and move them around.

## def•i•ni•tion

The word **congruent** is used when two figures are both the same shape and the same size. There's a bit of wiggle room when it comes to being the same shape: mirror images are okay, and it's also fine if the shape is pointing in a different direction. However, there's no leeway on being the same size. The sizes have to match exactly.

One of the tasks that you'll be faced with in a geometry class is to prove that certain pairs of triangles are congruent based on information about the measures of their angles and the lengths of their sides. But don't fret about that now. At this point, you'll just be asked to eyeball it and to guess when two shapes are congruent. It probably doesn't appeal to the perfectionists out there, but it's a welcome respite from rigor for those of you who are more big-picture people.

## Practice Makes Perfect

Problem 1: Draw two squares with equal side lengths. Are they congruent?

(!) Think twice before you cut any image of Marvin out of this book!

# It's a Shape of a Different Size

If shapes aren't congruent, there's still hope for them to show some lesser kind of sameness. If they aren't exactly the same, they might still be *similar*. You probably won't be shocked to hear that in math "similar" has a very specific meaning, and if you use the familiar everyday meaning of the word, you're going to make mistakes. Two shapes are similar if they have the same basic shape but are different sizes. Again, mirror images are okay, and it's still fine if they point in different directions. If you look at two squares, one may be big, the other little, but they are still similar because they have the same basic shape. You will find the same thing with equilateral triangles and circles. The idea is that they are the same shape, even if they aren't exactly the same. They are not all congruent, but they have a certain amount of sameness to them. All corresponding angles are the same and the ratios of the various sides and curves are also the same.

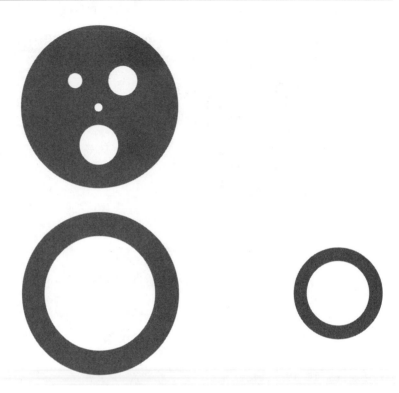

When you have two versions of the same shape in different sizes, this is also known as a dilation. Dilation is the process of taking a shape and scaling it up or down to get a similar version of itself. An analogy that works well here is with the pupil in your eye. When your pupil dilates as the amount of light lessens, it gets bigger but it stays the same shape (round). That's what's going on in this section. A shape will get bigger (or smaller), and you'll be calculating the size of the shape based on what you are told about the size of the original and the scale factor between the two different versions.

## def•i•ni•tion

Two shapes are **similar** if one is a scaled drawing of the other. That is, all of the corresponding angles and the ratio of sides are the same. Rotation or reflection is okay as well.

# Scale Factors and Resizing

When you have two shapes that are similar, you can make comparisons between the two different versions. You can look at a feature, such as the length of a side, from the large version and then calculate the length of the corresponding side on the smaller version.

This situation comes up when you have an image of a given size and you want to print it at a different size, a 5 × 7 to an 8 × 10. You might wonder where math comes into it. If you have photographs on your camera and you go to a drugstore or a discount store to print them, you just push the button for the size that you want. If you're working with images on your computer, you just drag the corner of the picture to make it bigger or smaller.

But what if you're standing in front of a photocopier and need to scale a document up or down? Maybe a book has printed a scaled-down version of the pattern for something that you want to make, and you want to make it in a different size? In any of these cases, you'll need to decide on the scale factor to enlarge or reduce the image that you're working with. In all of these problems you will need to multiply by a scale factor. This will either scale a small image up into a larger image or take a large image and scale it down to something smaller. If the scale factor is a number larger than 1, then you are scaling things up. If, however, you are multiplying by a number between 0 and 1, your scale factor is shrinking down your image into something smaller. Those are really your only options for a scale factor. If you multiply by 1 exactly, then that leaves your image unchanged. If your photocopier is set to copy at 100 percent, then your copy will be the same size as your original. Scale factors can never be negative. There's just no way for that to make sense.

There are a few basic types of problems of this sort that you're likely to encounter:

◆ You're given the scale factor and the size of the original, and you're asked to find the size of the reproduction.

◆ You're given the size of the original and the size of the reproduction, and you're asked to find the value of the scale factor.

◆ You're given some information about the original and some information about the reproduction. You're asked to find some other information about the reproduction. In order to accomplish this task, you need to find the scale factor as part of an intermediate step.

Clearly, the third of these is the sneakiest, as every time there's a problem with a hidden step in the middle, you need to be on your guard. So let's start with the simpler cases.

## Using the Scale Factor

If you're given the scale factor and the size of the original, then it's really easy to find the size of the copy. You just multiply. That's all there is to it—most of the time.

**Example:** You have an equilateral triangle, and each of its sides is 5 cm long. You enlarge it by a scale factor of 1.2. How long is each side of the larger triangle?

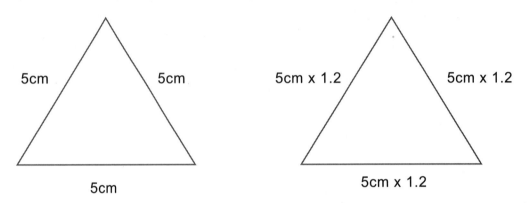

**Solution:** You multiply 5 cm × 1.2 = 6 cm. The scaled-up version of the equilateral triangle has sides that are 6 cm long.

**Example:** You have a square, and its sides are all 8 cm long. You reproduce it at 125%. How large is the copy of the square?

**Solution:** In this example, the scale factor is given as a percentage. Never fear, this is just a number. You remember how to rewrite a percentage as a decimal, right? (If not, go back and look at Chapter 5.) 125 percent is the same thing as 1.25, so the scale factor is 1.25. You calculate 8 cm × 1.25 = 10 cm, so the new square has sides that are 10 cm long.

Both of those examples had scale factors bigger than 1, meaning your copy will be larger than your original. But what if you're looking at a situation where you're shrinking something down? How do we fit Rhode Island onto a reasonably sized piece of paper? Before tackling that, let's look at a simpler example, say, a rectangle.

**Example:** A rectangle has sides of length 12 inches and 16 inches. It's duplicated with a scale factor of 0.75. How large is the copy of the rectangle?

**Solution:** Here we have two sides to deal with, so we work with them separately. To scale the 12-inch side, we multiply it by 0.75, and 12 inches × 0.75 = 9 inches. Then we can deal with the 16-inch side, multiplying it by 0.75 to get 16 inches × 0.75 = 12 inches. Therefore, the new rectangle measures 9 inches by 12 inches.

## Calculating the Scale Factor

So where does this scale factor come from? Does it float down on a cloud, ready for you to use it when you need it? Maybe you just snap your fingers and it appears, waiting to be multiplied into the problem. Alas, no, the answer is much more prosaic: division.

You need to be given a pair of corresponding lengths, one from the original and one from the scaled copy. It's not that big a deal if you're working with a square or an equilateral triangle or any other regular polygon. When all the sides are the same length, there's no need to worry about the long side or the short side or specifying any of them in a special way. Where the trouble comes in is when you have a general rectangle or some other shape where sides come in different lengths. In order to find the scale factor, you need a pair of sides that have the same role in both the original and the copy. If you have a rectangle, you'll want the short side in the original and the short side in the copy, or the long side in the original and the long side in the copy. You can't mix and match. You can't take the short side in one picture and the long side in the other, no matter how enticing the numbers are. It just doesn't work that way.

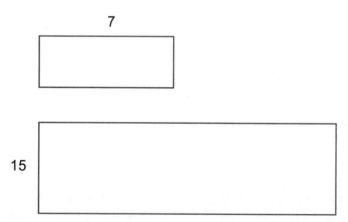

So once you have these corresponding sides, what do you do with them? It was hinted that division was involved, but you remember that order definitely does matter in division. It matters which side comes first in the division problem. So here's how it works: you take the length in the copy and divide it by the length in the original. That gives you your scale factor.

**Example:** You have ordered over the Internet a set of the secret plans for making a machine that lets you control other people's minds. Unfortunately for you, in order to get them to all fit in the book, they've scaled them down. If a part of the machine is drawn as being 6 cm long in the book and it needs to be 21 cm long in real life, what is the scale factor that you need to use to enlarge the plans?

**Solution:** Because the copy you're making is going to be larger than the original, you know the answer has to be a number larger than 1. If you get a number that isn't larger than 1, you know that you've made a mistake and you need to go back and check your work.

The rule for calculating the scale factor first says that you need to find corresponding sides. Fortunately for you, there is only one set of sides discussed in this problem, and the problem tells you that they're corresponding (they represent the same thing in the plans), so you don't have to worry about that.

All you need to do is to take the size in the copy and divide it by the size in the original. This gives you $21 \div 6 = 3.5$. This means that if you were photocopying the book of plans, you'd need to set the machine to 350 percent in order to print them at their intended size.

**Example:** Find the scale factor between the following rectangles.

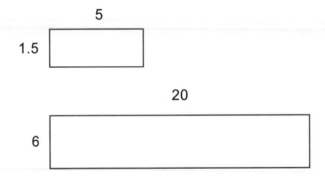

**Solution:** First find the corresponding measurements on the two figures. In this case, we have two sets of lengths that are already marked on both diagrams: the long side of the rectangles and the short side of the rectangles. Let's choose the long side so we don't have to deal with decimals.

The larger figure has the long side of the rectangle equal to 20 and the smaller figure has the long side of the rectangle equal to 5. So the scale factor between these two figures is $20 \div 5 = 4$.

**Practice Makes Perfect**

Problem 3: Find the scale factor between these two regular pentagons.

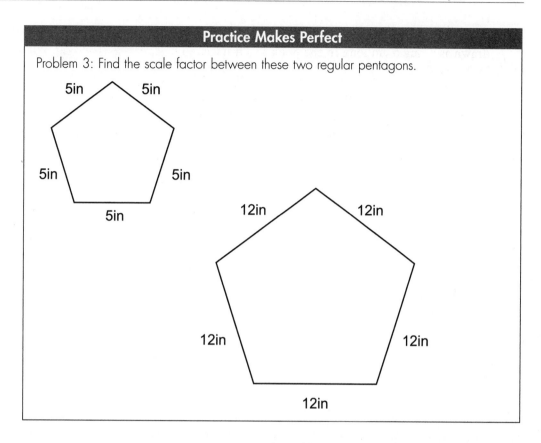

So now you're ready to handle the most complicated type of problem. In these you typically will be given several lengths, some from the original, some from the copy. And you'll be asked to find another length in the copy. There are two ways that you can do this type of problem. You can either set it up as a multi-step scale factor problem. Or you can set up an equation that uses proportions. Both methods (if used correctly) will give you the same answer. However, if you're enrolled in a course where the teacher is trying to get you to use a specific method to solve the problems, then it's entirely possible that only one of them would be considered the "right" way to solve the problem. Yes, you might think that it's unfair, but there is no point in debating the logic of how the right number can be considered a wrong answer. Better to spend your time learning how to use the method that your teacher prefers.

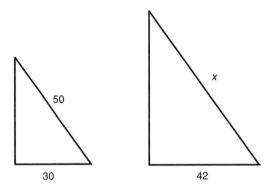

**Example:** In the similar triangles illustrated in the previous figure, find the length of the side labeled *x*.

**Solution:** If you knew the scale factor, this would be an easy problem. This would be like the ones that you saw at the beginning of this section. You would want to know what 50 scales up to in the larger triangle, so you would multiply 50 by the scale factor.

Unfortunately, you aren't given the scale factor, but you can find it. You know that the side that's 42 in the big triangle goes with the side that is 30 in the smaller triangle. Because you're trying to figure out stuff about the larger triangle, it's designated as the copy, and the smaller triangle is the original.

You calculate the scale factor by dividing the size of the copy by the size of the original. In this case it's 42 ÷ 30 = 1.4. Does it make sense for the scale factor to be greater than one? Sure it does, because you're scaling up from a smaller triangle to a larger one.

Now to finish the problem, you can multiply 50 by your shiny new scale factor. 50 × 1.4 = 70. This means that the side labeled *x* in the larger triangle is 70.

Now what if you wanted to use the fact that you already had something named *x* and you wanted to set up an equation? That's the sort of thing that you do in math class, especially when something is already given the name *x*. It almost calls out for you to do that. And math doesn't disappoint. You can set up a proportion and solve it to find the length.

When setting up a proportion, you'll be comparing things in corresponding pairs. There are two ways that you can do it: you can look at the ratios of sides within the same triangle or you can look at the ratios of corresponding sides from the different triangles. Don't read that to mean that you can set things up every which way without paying attention. It does mean that the number of right ways to set up the problem has a fighting chance against the number of wrong ways to set up the problem, but there are still a lot of wrong ways, and you don't want to get tangled up with them.

The first way of setting this up would be to look at the ratio long side of the big triangle over the short side of the big triangle and the ratio long side of the small triangle over the short side of the small triangle. That's fairly wordy. Put into math notation, the ratios that you're looking at are $x/42$ and $50/30$. The amazing thing here is that because these triangles are similar, these ratios are equal! So you can set up an equation $x/42 = 50/30$. To clear fractions and solve for $x$, you multiply both sides of the equation by 42. Multiplying everything out, you get that $x = 42 \times 50/30$, which equals 70. As promised, this gives you the same answer as before.

What we were really doing before in the problems with the scale factor was to set up a proportion that compared the two different triangles. Instead of having a ratio that compared one side of a triangle to another side of the same triangle, you can set up your ratios so that they compare the side of a triangle with the corresponding side of the other triangle.

**Amy's Answers**

It doesn't matter if you do long over short or short over long, just as long as you're consistent. The best way to set up this type of problem is any way that you can avoid having variables in the denominator.

So this same problem can be set up another way illustrating that. (I know, I know, enough of this problem already. I can assure you that it is as sick of you as you are of it.) To do it this way, you'd line up the ratios as short side of big triangle over short side of small triangle and long side of big triangle over long side of small triangle. In this case it would be $42/30$ and $x/50$. Again, because of the magic of similar triangles, these are equal, so $x/50 = 42/30$. When you solve the equation by multiplying both sides of the equation by 50, you get $x = \dfrac{(50 \times 42)}{30}$ (doesn't that look familiar), which is still equal to 70.

| Practice Makes Perfect |
| --- |

Problem 4: Find the length of the side labeled x.

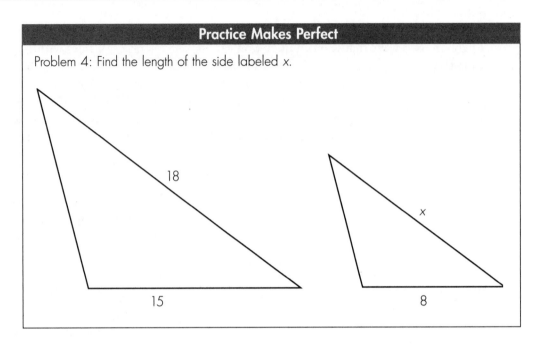

# Change in Areas

By now you should be thoroughly indoctrinated with the idea that to scale the lengths in an image up or down, you multiply by the scale factor. So what do you think happens when you compare the area of the enlarged (or reduced) image with that of the original?

There's a hard way and an easy way.

Let's get the hard stuff out of the way. You can take the time to find all the necessary lengths of sides in the new figure, using the scale factor, and then use the area formulas to find the area of the new shape. This always will work. Sometimes it is tedious. Sometimes this is how your teacher will want you to do it, even though there is an easier way.

The easy way? You multiply the original area by the scale factor squared. Really, that's it. That's why the scale factor is so wonderful and we love it so much!

**Example:** Find the area of the larger of the two following similar triangles.

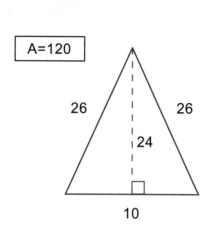

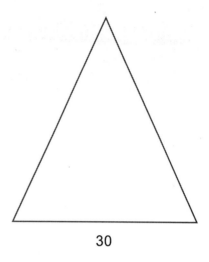

**Solution:** The easy way—The base of the smaller triangle is 10, and the base of the larger triangle is 30. Thus, the scale factor is 30 ÷ 10 = 3. We could calculate the area of the smaller triangle using the formula for area; however, the figure is already helpfully labeled, telling us that the area is 120. Since area changes as the square of the scale factor, the area of the larger figure is $120 \times 3^2 = 1,080$.

The hard way—The base of the smaller triangle is 10, and the base of the larger triangle is 30. Thus, the scale factor is 30 ÷ 10 = 3. Now we have to calculate the other measurements by multiplying by the scale factor. $26 \times 3 = 78$ and $24 \times 3 = 72$. We can look at 72 as the height of the triangle with a base of 30 and use the area formula to see that the area is $\frac{1}{2} \times 30 \times 72 = 1,080$, as before.

The "hard way" may not have seemed too difficult in this case, but sometimes it's a real pain to have to convert every measurement and find the areas with the new, scaled values.

**Example:** Find the area of the smaller figure.

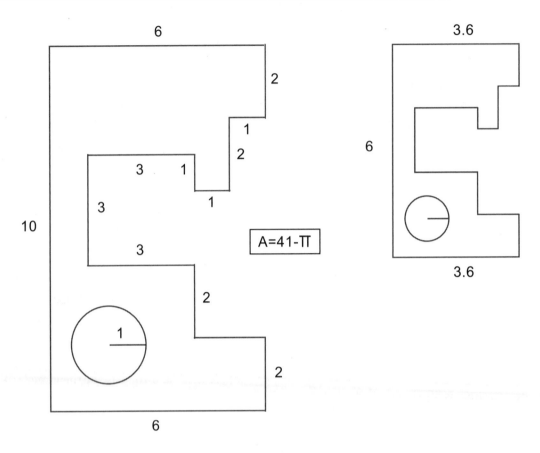

**Solution:** If you'd like to convert all of the measurements in the larger figure to find them in the smaller figure, be my guest! As we're given the area of the larger figure as 41 − π and we can find the scale factor by comparing measurements in the figure, we can find the smaller area by just multiplying a few items together.

First, we find the scale factor by picking a pair of corresponding sides and dividing the length in the copy by the length in the original. Probably the easiest side is the one on the left of both shapes; it's 10 units long in the original, and 6 units long in the copy. Calculate the scale factor as 6 ÷ 10 = 6⁄10. The smaller area is equal to the larger area multiplied by the proper scale factor squared. So

$$\left(41-\pi\right)\left(\frac{6}{10}\right)^2 = \frac{\left(41-\pi\right)36}{100}$$

is the area of the smaller figure. Could we convert that to a decimal? Sure, we could, but it's better to leave the answer exact unless someone tells you otherwise. When putting a number that ugly into your calculator, you're likely to make a mistake.

## The Least You Need to Know

- Two shapes are congruent if you can cut one out and put it on top of the other; a perfect match!

- Two shapes are similar if you can shrink or enlarge one of the figures until it's congruent to the other.

- To convert between lengths in two similar figures, multiply the original length by the scale factor.

- To convert between areas in two similar figures, multiply the original area by the scale factor squared.

- Remember to look at the scale of any map before using it as a guide for your travels.

# Part 6

## Data Analysis

We live in a world of data. Studies are being done all the time. Sports teams want to track the progress of their players. Savvy shoppers want to track the prices at the grocery store. How do we make sense of all the numbers that we're bombarded with every day? Statisticians have come up with great ways to do that and also with methods to understand the story that the data tell us. In this part, you'll learn how to take numbers and make them tell a story.

*"I overheard Farmer say that there's a .001% chance that an asteroid is going to hit the barn."*

# Chapter 18

# Collecting Data

## In This Chapter

- ◆ Gathering data
- ◆ Drawing pictures
- ◆ Calculating measures of central tendency

One of my colleagues says, "Statistics don't lie. Statisticians do." Actually, most of my colleagues say that. But most of my colleagues are statisticians, so they're probably lying. Unless you learn the language and methods of data and statistics, you will be the one that they are lying to.

The first step toward understanding statistics and statistical reasoning is to become familiar with data. In this chapter you learn about data and how it's collected, represented, and analyzed. You might even learn how to lie to people yourself!

## What Is Data?

Any time you have a collection of numbers or characteristics that describe something, you have data. These could be the scores on an exam, the shoe sizes of the people in your family, the number of people that each of your pets has bitten, or the color of the bites a week after the fact. Data also is collected when scientists run experiments. How did the people in the treatment group compare to people who didn't get the treatment? Hopefully they're less likely to be dead.

# Why Bother?

You might think that the major reason for collecting all of these numbers and then issuing them to you for counting, analysis, and graphing is to waste your time and to give you something to do to keep you off the streets. This is not necessarily the case. Data analysis is big business, and you need to understand it to make sense of the information that you're being told.

Frequently the results of studies are reported in the paper without much analysis, and if you can't understand data, you won't know what's going on or whether it is accurate. You might find it acceptable to be blissfully ignorant of current events, but remember that the results of medical tests are also reported in terms of statistical information. Being unable to understand this information is not merely annoying; it could be dangerous to you or those you love.

There are several reasons researchers might collect data:

♦ Information might be collected from everyone in a group (such as everyone's scores on an exam) to come up with a way to describe the attributes of the group.

♦ Information might be collected from some people in a group and used to make inferences about the entire group. You might ask some people in a school what their favorite ice cream flavor is and assume that the most popular flavor among the people you ask is the most popular flavor for everyone in the entire school.

♦ Scientists collect data from several different groups of experimental subjects. These could be patients in a drug study, cells in a laboratory, or some other scientific situation such as how many molecules of water are at a water park. By comparing the groups, the scientists can make inferences about the effectiveness of a treatment.

# Qualitative vs. Quantitative

When you're describing something, you have a lot of choices of how to talk about it. Let's say that you were going to try to describe your math teacher. You can describe features that can be measured or counted. For example, your teacher's height, weight, and number of years teaching are all things that can be described by numbers. But things like niceness and helpfulness are qualities that are hard to measure. You can't just point a nice-o-meter at someone and have it spit out a number everyone will agree with.

This breakdown between things that can be measured and reported as a number and things that can't is the distinction between *quantitative* and *qualitative* data. Anything that can be described with a number (a quantity) is quantitative data, while anything else (a quality) is qualitative data.

When you have quantitative data, you can apply all sorts of statistical tests to report its features. One of these is the range. The range is simply the difference between the largest value and the smallest value.

**Example:** The scores on an exam are: 68, 73, 78, 79, 80, 83, 84, 92, and 96. What is the range of the data?

**Solution:** The range is the largest value minus the smallest. In this case it would be 96 – 68 = 28.

| **Practice Makes Perfect** |
| --- |
| Problem 1: The ages of the children in the Snook family are 2, 17, 10, 4, and 5. What is the range of their ages? |

# Recording Data in a Frequency Table

Once you have all this data, what are you going to do with it? Ideally you would have asked and answered that question before you even started collecting the data. But now you have it, and you're trying to use it to answer some questions. What should you do?

If you only have a little bit of data, like if you asked everyone in an elevator what floor they were going to, you probably wouldn't need to do much with the data. By the time you started to run statistical tests, probably everyone would have gotten off the elevator, rendering your data less useful. On the other hand, if you had a lifetime of collected information about your favorite ballplayer, you'd need to do something with this data to make sense of it.

Let's think about the situation of an exam given to all of the students in a school. More and more testing happens in schools, and sooner or later, everyone is going to have to take some sort of standardized test.

A frequency table is great for answering questions that begin "how many." If you want to know how many people scored in a certain range, how many students got above a specific score, or how many students were below a given threshold, a frequency table is going to be able to give you the information that you're looking for.

It's pretty easy to make a basic frequency table. You make a list of all the values that appear in your data, and you count how many times each of them appears on the list. If you want to be extra-fancy, then you also can say what percent of time each value or range appears.

**Example:** You have collected information about the shoe sizes of the students in your math class. You get the following data: 6, 6.5, 6.5, 7, 7, 7.5, 7.5, 7.5, 7.5, 8, 8, 8, 9, 9.5, 10, 10, 10.5, 10.5, 10.5, 11, 11, 11.5.

**Solution:**

| Value | Frequency | Percentage |
|-------|-----------|------------|
| 6 | 1 | 4.5% |
| 6.5 | 2 | 9.1% |
| 7 | 2 | 9.1% |
| 7.5 | 4 | 18.2% |
| 8 | 3 | 13.6% |
| 9 | 1 | 4.5% |
| 9.5 | 1 | 4.5% |
| 10 | 2 | 9.1% |
| 10.5 | 3 | 13.6% |
| 11 | 2 | 9.1% |
| 11.5 | 1 | 4.5% |

To find the percentages, you take the frequency and you divide it by the total number of values, then convert that decimal to a percent. In this case, there are 22 values, so the percentages are found by taking the frequency of each value and dividing by 22, the total number of measurements.

You can make frequency tables from qualitative data, too.

**Example:** You watch the cars driving by your house and you record what colors they are. Make a frequency table from the data: white, red, blue, silver, silver, white, green, grey, silver, red, silver, blue.

**Solution:**

| Value | Frequency | Percentage |
|-------|-----------|------------|
| white | 2 | 16.7% |
| red | 2 | 16.7% |
| blue | 2 | 16.7% |
| silver | 4 | 33.3% |
| green | 1 | 8.3% |
| gray | 1 | 8.3% |

This is all well and good, but sometimes you have a lot of data and maybe each individual value only shows up once or twice. It seems like a frequency table wouldn't be much help in this situation. But it can be— if you aggregate your data into sensible groups. Then you can count how many times something from each group shows up. For example, you could rearrange your car color data into a group of colorful colors and shades, another name for grayish colors. White, silver, and gray make one group, and red, blue, and green make another.

> **Practice Makes Perfect**
>
> Problem 2: Make a frequency table of the eye color of students in your class. Their eye colors are: brown, blue, green, brown, grey, green, hazel, green, blue, blue, brown, green.

| Value | Frequency | Percentage |
|-------|-----------|------------|
| colorful | 5 | 58.3% |
| shadeful | 7 | 41.7% |

# Sorting Data

You probably wouldn't want to make a frequency table showing the frequencies of each of the grades earned by all of the students in a school district who take a standardized test. Really, it doesn't matter that much how many 83s there are and how many 84s there are. The school is probably more interested in how well students did in general. They probably want to know how many scores were in the 90s or how many scores were below a certain threshold.

# Bins

When we sort data, we say that we sort it into bins. These are not physical objects that hold the data but rather a mathematical metaphor of sorts. A bin is simply a range of values that the data could take on. For example, if you're looking at test scores, you might want to know how many students scored from 80 to 89 on the exam. Depending on your data and what you want to say about it, you will need to decide what bins make sense. If you're lucky and people are making life easy for you, you won't have a choice. The problem will just tell you what to do.

Suppose you have a radar gun to see how fast the cars are going by your house and you end up with the following values:

Car 1:        29 mph

Car 2:        31 mph

Car 3:        25 mph

Car 4:        24 mph

Car 5:        23 mph

Car 6:        35 mph

Car 7:        33 mph

Car 8:        20 mph

Car 9:        37 mph

Car 10:       39 mph

Car 11:       22 mph

Car 12:       38 mph

No two cars are going the same speed, so making a frequency table seems a little silly. If the speed limit outside your house is 30 mph, it might make more sense to count how many people are obeying the speed limit and how many people are breaking it. To do this, make bins for speeds at or below 30 mph and speeds above 30 mph.

| Value | Frequency | Percentage |
| --- | --- | --- |
| 0-30 mph | 6 | 50.0% |
| 31+ mph | 6 | 50.0% |

Now you can say something real with these statistics: about half of the cars that pass your house are speeding!

---

### Practice Makes Perfect

Problem 3: The scores on an exam are: 68, 73, 55, 78, 83, 79, 80, 83, 84, 92, and 96. If 80-89 is a B, how many students got a B or better?

Problem 4: Take another look at the previous data from cars passing your house. How many were driving within 5 miles per hour of the speed limit of 30 mph?

---

# Outliers

Think back to the recurring example of grades on a test. (Yes, I could come up with a new example, but this one has been working so well that I'm reluctant to give it up.) You know how sometimes most of the class will score about the same thing, maybe in the 80s, and then there is one know-it-all who scores 100? Or it could happen the other way around. You might have taken a really, really, really easy test and everyone got an A, except for one person who must have slept through it and who failed. In these situations it would be relatively easy to describe what's happening—if it weren't for that one weird value.

You also can think about this in terms of the weather. Imagine that it's wintertime somewhere in the northeastern United States (or southeastern Canada) or somewhere else where it is usually cold and snowy in the winter. Suppose you collect temperature data every day for the month of January. Almost every day is around freezing, except for one day when it is really, really warm. Unseasonably warm. If it weren't for this one warm day, you could say that it's cold in January.

These isolated weird values that are either much larger or much smaller than all the other values in the dataset are called outliers.

How weird do they need to be in order to be considered an outlier? There's no fixed rule for this. In medical data the outliers are often important. Even if 99 out of 100 people who take a new medicine get over their cold faster and have no side effects, we can't ignore the fact that 1 out of 100 people die from it. On the other hand, if you measure a car on your street moving at 0 mph because it's parked, it's probably reasonable to ignore it if you're trying to find out how many cars are speeding on your street. The important thing is to remember that when someone removes an outlier, they're removing a data point that doesn't fit with the rest of the data.

Sometimes when you see a list of data, the outliers will be marked with asterisks. This tells you that they're outliers and that they may be excluded from any calculations that are based on the data.

# Displaying Data

Once you have your data all collected and corralled into bins, the next step is to decide what to do with it. Working statisticians have all sorts of mathematical and computational tools available to them to wrangle the data and squeeze all the information out of it. Our goals are a bit more modest. We'd like to share the information that's been collected and to describe the situation the data represent.

One of the most effective ways to communicate data is with a picture. Just like the expression that a picture is worth a thousand words, a visual display of your data can show a lot of information about the system that you're measuring. Depending on what you're trying to convey, there are different ways to make a visual display of your data.

## Pictograph

In a pictograph you have a picture stand for some stuff. For example, if you were an officer in a club that sold oranges as a fundraiser and you were making a pictogram representing the annual sales of oranges, you might let a picture of an orange stand for a certain number of cases of oranges sold. How many cases should each orange picture stand for? That depends on how many your club sold. The whole point of the pictograph is to make the picture of the data fit nicely on the page. If you sold a lot of oranges, each picture should stand for a lot of cases. If you didn't sell very many, then each picture might stand for one case. The idea here is that you want the diagram to look nice on the page, and you don't want your reader to have to count a zillion oranges that are all smooshed together on the page. You also don't want one massive orange for all the rows. We're looking for a happy medium here.

---

**Practice Makes Perfect**

Problem 5: Make a pictograph of the eye colors from Problem 2. You should already have the frequency table made!

---

You're the head of the local ice-cream cartel, and you want to show your supporters the evidence of some other sinister organization gradually taking over ice cream sales during the fall, dominating them during the winter, but retreating as the summer begins. The following table shows your sales data.

| Month | Sales |
|---|---|
| August | 1000 gallons |
| September | 750 gallons |
| October | 400 gallons |
| November | 200 gallons |
| December | 150 gallons |
| January | 100 gallons |
| February | 90 gallons |
| March | 110 gallons |
| April | 170 gallons |
| May | 500 gallons |
| June | 750 gallons |
| July | 1200 gallons |

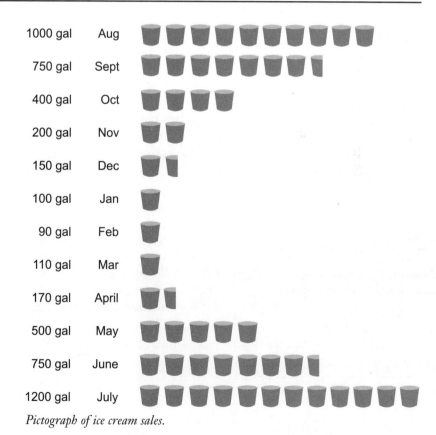

*Pictograph of ice cream sales.*

Based on this sales data, our pictograph might look like the previous figure. Don't forget to write down what your scale is!

# Histogram

The histogram possibly might be the gold standard of data displays. Every number-crunching piece of software out there is eager to make a histogram for you. A histogram is simply a graphical representation of a frequency table. The horizontal axis of your histogram will have spaces for the bins that you sorted the data into. The vertical axis will measure the frequency for each bin. Each bin will be represented by a bar whose height is the same as its frequency. Data that appear more frequently will have longer bars. Data that appear less frequently will have shorter bars. One bit of histogram etiquette: the bars have to be drawn so that they touch each other on their sides.

In order to make a histogram, you need to start with a frequency table. You can't even think about starting to draw your histogram until you have a frequency table. Once you have the frequency table, you start by making your axes. Instead of crossing in the center like the coordinate axes from algebra, these axes touch in the lower left, making an L-shape. You will label your horizontal axis by whatever your data is measuring, and your vertical axis will be frequency. Mark intervals on the horizontal axis that describe your bins from your frequency table; the vertical axis is marked off with the range of frequencies you observed. For each bin, you make a bar whose height represents the frequency that you found in your frequency table. And don't forget: the bars have to touch each other.

We can rewrite our pictogram from the ice-cream cartel as a histogram like this.

---

**Practice Makes Perfect**

Problem 6: Make a histogram of the data for speeds of cars from the section about 'Bins.' Since no two cars are traveling the same speed, each bin should be a range of numbers. The ranges should be: 22-23, 24-25, 26-27, and so on until you cover the entire range of your data.

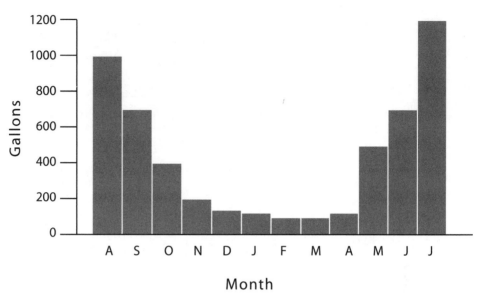

*Histogram of ice cream sales.*

## Bar Graph

Bar graphs look a lot like histograms and have many of the same features in the way they look and feel. The difference is that each of the column bars stand for different measurements of different items rather than a frequency table. You could compare your yearly ice cream sales to that of Raskin Bobbins and Stold Cone using a bar graph. If Raskin Bobbins sells 1400 gallons of ice cream over the year and Stold Cone sells 2500 gallons, a bar graph of yearly sales would look like this.

> **Practice Makes Perfect**
>
> Problem 7: Make a bar graph of the eye color data from Practice Makes Perfect Problem 2. You should already have a frequency table made! Compare the bar graph to your pictograph from Problem 5.

## Stem-and-Leaf Plot

Don't get your hopes up about the chance to make a lovely diagram representing a tree. While the stem-and-leaf plot has a pretty name that makes you think of lush foliage, it's actually a fairly routine statistical plot. One thing that it has working to its advantage is that it's really easy to make and it displays a lot of information about your data. It's sort of a combination between a histogram and a complete list of the data.

With most stem-and-leaf plots, there is one major drawback. You need to group your data into bins in a very specific way. Every datum in a bin needs to begin the same way; typically this means that everything in a bin has the same first digit. This is somewhat awkward if your data is wildly spread or if it is naturally clumpy in a way that doesn't work well with grouping things together by first digit.

Where a stem-and-leaf plot works phenomenally well is in showing the grades on an exam. In my class, any score in the 90s is an A. Scores in the 80s are Bs, and grades in the 70s are Cs. Ds are in the 60s, and everything below 60 is an F. So when I report the scores for an exam, the bins come out pretty nice if I use letter scores. The stem-and-leaf plot was practically made for this situation.

So how do you make this plot? As I said, it's really easy. Along the left side of the table, you list all of the ways that a grade could begin—everything up to (but excluding) the units digit of the number. So for a plot that represents grades on an exam where everyone earned between a 50 (begins with 5) and a 100 (begins with 10), the setup would look something like:

10|

9|

8|

7|

6|

5|

Now all you do is take every score, and you write its units digit in the appropriate row. If someone scored a 72 on my exam I would take the row 7| and I would write 7|2. If someone else got a 73, this row would become 7|23. The 2 is from the 72, and the 3 is from the 73.

Imagine that the grades on the exam were: 51, 59, 61, 64, 64, 66, 68, 72, 73, 76, 76, 77, 78, 82, 84, 85, 86, 88, 92, 97, 100.

Then the stem-and-leaf plot would look like:

10|0

9|27

8|24568

7|236678

6|14468

5|19

> **Practice Makes Perfect**
>
> Problem 8: Think of ten different people you know whose ages you also know. Make a stem-and-leaf plot of their ages.

A completed stem-and-leaf plot looks like a histogram only sideways. You can tell from this graph that more students scored in the 70s than in any other range. You also have all of the scores listed. If you were in the class and wanted to calculate more statistics on the exam data, you could just read it off.

## Venn Diagrams

Some people have only cats, others only dogs, and there are individuals who have both. If you make a bar plot of how many people have cats and how many have dogs, some of the important information about pet owners will be missing because you can't figure out how many people own both! You sometimes can solve this problem by making something called a Venn diagram.

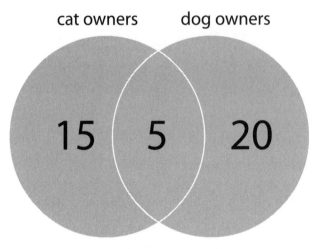

*A Venn diagram.*

Venn diagrams are pictures that can help show how many people fall into multiple categories of some survey, such as owning cats and dogs. There are several shapes overlapping, and being inside a shape means you have a characteristic assigned to that shape. The number inside each little area is how many fall into all the categories represented by the shapes enclosing the area. Blah blah blah. Let's look at an example.

**Example:** Some people only have a car or only have a bike, but others have both. If you collect data from your classmates and find that 35 have only cars, 20 have only bikes, and 45 have both, how would you draw the Venn diagram?

**Solution:** The corresponding Venn diagram will have a "car" area with 35 in it, a "bike" area with 20 in it, and an overlapping area with a 45 in it. It should look something like the following:

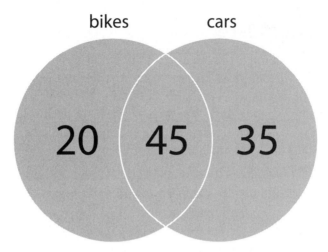

*A Venn diagram of the car area, bike area, and overlapping area.*

If many of your data belong to multiple groups, that should set off an alarm telling you to use a Venn diagram to represent your data. It doesn't make much sense to have a Venn diagram of the color of a car since most cars are only one color—none of the shapes would overlap—but you should use Venn diagrams to describe what genres of music your classmates listen to as most of them probably like more than one genre.

---

**Practice Makes Perfect**

Problem 9: Johnny Doh took a poll of his friends and found out that 16 of them liked only vanilla ice cream, 2 of them liked only chocolate ice cream, and 24 of them like both chocolate and vanilla ice cream. Make a Venn Diagram representing Johnny's data.

---

# Line Plot

One of your neighbors makes $2 per day at work, and she will make about $700 per year. Another of your neighbors makes $5 an hour mowing lawns for a total of $2,000 per year. A third neighbor makes $40,000 per year. The first is probably a little girl selling lemonade by the side of the road, the second a boy who wants some pocket money, and the third might be a manager who has been working with the same company for several years. In order to understand these and other data about salary, it's often necessary to look at the age of the individuals as well as the amount of money they make. If a 25-year-old were making $200 per year, we'd probably be fairly surprised. We wouldn't be as surprised if a 12- to 15-year-old made $200 in a year, though.

# Just an Average Joe

Sometimes you're looking for one number to describe an entire situation. It might be something like the salaries of the graduates of a certain school. The school might claim that most of its graduates go on to do well in the world, so it must be a good school. It might be something like the price of gas in an entire city. A shoe store might want to know what the most popular shoe size is. A teacher might want to describe how well a class did on an exam. In all of these cases, a complicated situation is described by one number.

Describing complex and nuanced goings-on with a single number is not without its perils. Should you choose the highest number? The lowest number? One in the middle? Something else? There are several choices you can make, and depending on the situation, some choices are better than others. Three classic ways to pick a number are the mean, the median, and the mode. These are often called measures of central tendency.

## Mean

Despite its name, you don't need to be afraid of the mean. It's not going to talk about you behind your back, make fun of you in front of everyone, or beat you up. The mean is just another name for the average. Doesn't average seem less threatening than mean?

The mean gives you a sense of the typical value for the situation. When we calculate the mean grade on an exam (often called the class average), we're finding a score that represents, in some sense, how the typical student did on the exam. The mean salary for an industry would tell you how much the typical worker earns. The mean number of cars owned by families in a certain city gives you an idea of the typical transportation patterns in that area.

Of the three measures of central tendency, the mean is the only one that you're likely to need your calculator for; fortunately it requires little more than adding and dividing. To find the mean of a set of numbers, you add them all together and then divide by how many numbers are on the list.

**Example:** Find the mean of 4, 8, 10, 3, and 9.

**Solution:** First off, don't worry that the list is not written in order. That doesn't make a bit of difference when you're working with the mean.

To calculate the mean, you add the numbers together and then divide by how many numbers are on the list. In this case, you would be dividing by 5 because there are five numbers on the list.

You compute $(4 + 8 + 10 + 3 + 9) \div 5 = 6.8$.

**Example:** I gave an exam the other day to my class, and the grades were: 49, 83, 91, 94, 95, 96, 96, 98, 98, 98, 98, 99, 99, 100, 100, 100, 100. What was the mean score earned on the test?

**Solution:** This is the same type of problem: you add up how many grades are on the list, in this case 17. Then, you add up all the scores and divide by 17.

$(49 + 83 + 91 + 94 + 95 + 96 + 96 + 98 + 98 + 98 + 98 + 99 + 99 + 100 + 100 + 100 + 100) \div 17 = 1594 \div 17 = 93.76$.

Therefore the average grade on the exam was a 93.76.

The mean is sometimes misleading. I know, you probably thought that there was a reason that we called it mean. I don't think it's deceptive on purpose; it's just that when your data has some outliers, the mean is easily swayed to one side or the other.

Imagine the case of a school trying to describe the success of its graduates by publicizing the average salary earned by all of the students that ever graduated from that school. They would take the annual incomes of everyone who had graduated, add those all up, and then divide by the number of graduates.

For many schools this would give a decent sense of what happens to the alumni. Do they go on to do great things? Or do most of them end up asking, "Do you want fries with that?"

The problem comes up in some schools whose graduates include some outliers. Take, for example, the high school that Bill Gates graduated from. (He might have dropped out of Harvard, but he definitely finished high school!) Because Bill Gates makes a superhuman amount of money (about a kajillion gazillion dollars), you would be adding together a whole bunch of normal people's salaries together with the insane wealth of Bill Gates. Even after you divide by the number of alumni, you'll still have a very big number. While I'm sure that Bill's classmates are doing okay for themselves, averaging in Bill Gates's money gives a misleading value for the mean.

You see the same phenomenon in the test grades. The student who scored a 49 on the exam is bringing down the class average. If you leave out that score, the mean goes up to 96.56.

The moral of the story is that the mean is a pretty good measure of what's typical most of the time, but if you have extreme outliers, it becomes much less useful.

---

### Practice Makes Perfect

Problem 10: Carl, Janet, and six of their friends timed themselves to see how long it took to run 100 meters. Carl fell asleep at the start line for a few minutes and Janet secretly wore shoes with wheels in the bottoms. The times in seconds for the six friends were: 14, 16, 16, 16, 20, and 21. Carl took 303 seconds to wake up and run it, and Janet took only 5 seconds to roll 100 meters. Calculate the mean of all eight times, then calculate the mean without the outlier times of Carl and Janet and compare the two means.

---

# Median

The median is another measure of central tendency, and it is not as fazed by outliers as the mean is. Another advantage of the median is that you often don't have to calculate anything in order to find it. Be careful, though; when you're looking for the median, you have to start by making sure that the numbers on your list are in order. They can go from smallest to biggest or from biggest to smallest. All that matters is that they don't jump around at all.

### Kositsky's Cautions

Be absolutely certain that the list is in order. If you accidentally have numbers where they shouldn't be, you'll have to start all over again or your answer will probably be wrong.

So how do we find a number without calculating? Do we just pull it out of the list? In some ways, yes. The median is the middle number on the list. It's the number such that there are just as many numbers before it as there are after it.

Here's a great way to find the median. Start by putting your list in order (this is absolutely essential). Then put one finger on the first number on the list and one on the last number on the list. Then move each finger one number toward the center. Keep doing this until they both land on the number in the middle of the list. This method always works if there is an odd number of items on the list, like in the following example.

**Example:** Find the median of 4, 8, 2, 7, 7, 4, 9.

**Solution:** Start by putting the list in order: 2, 4, 4, 7, 7, 8, 9. Put one finger on the 2 and another on the 9, and start moving in toward the middle one number at a time. You'll end up with both fingers on a 7. Therefore, the median is 7.

So what happens if there is an even number of items on the list and there is no middle number? In that case, we take the two numbers closest to the middle of the list and take the average of the two.

**Example:** Find the median of 8, 5, 9, 15, 3, 13, 1, 10.

**Solution:** Start by putting the list in order: 1, 3, 5, 8, 9, 10, 13, 15. If you start at the ends of the list and count in toward the middle, you will end up with your fingers on 8 and 9. This means that the median will be the average of 8 and 9 which is $(8 + 9) \div 2 = 8.5$. Therefore, the median is 8.5.

> **Practice Makes Perfect**
>
> Problem 11: Find the median of the following data: 13, 2, 8, 9, 10, 16.

# Mode

When politicians speak of "The Common People," they probably are referring to the *mode* of the population—the most commonly occurring people. The mode is the most common value in a dataset. It's whatever shows up the most. That's it. The mode is a different kind of measure of central tendency than the other two because it doesn't have anything to do with the numerical middle of the list. The median is clearly at the middle, and you can make a really good argument for the mean being in (or at least near) the middle, but it's hard to say that about the mode.

One advantage of the mode is that you can use it to talk about the *average* car color. If you try to find the median car color, what ordering of the colors will you use? If you try to find the mean car color by averaging all of them you'll probably just come up with ugly. The mode of car colors is a meaningful result that indeed describes something that could be considered the average car color. If you have 3 red cars, 1 blue, and 1 white, the mode of the color of cars is red because the majority of cars are red.

**Example:** Find the mode of 7, 20, 7, 8, 5, 13, 7, 8, 12, 9.

**Solution:** The number that appears most frequently is 7, so the mode is 7.

> **Practice Makes Perfect**
>
> Problem 12: The students in Rosie's class took a poll of how many pets each student had at home. The data was: 0, 0, 0, 0, 1, 1, 2, 3, 2, 5, 2, 6, 4, 5, 0, 1, 1. Find the mode of pets that students have in Rosie's class.

## The Least You Need to Know

◆ Sorting data into a frequency table is the first step in choosing which way to represent it.

◆ There are many ways of looking at data, including Venn diagrams, histograms, bar plots, and pictographs.

◆ The mean, or average, is what you get when you add together all the data and then divide by the number of data points.

◆ The median is the middle value in a dataset.

◆ The mode is the most common value in a dataset.

# Chapter 19

# Probably Probability

## In This Chapter

- ◆ Dependent events
- ◆ Independent events
- ◆ Card games

The fields of probability and statistics go together like peanut butter and jelly. Sure, you can have each of them separately, but there is something special about putting them together.

Statistics can be used to describe data, but they also can be used to tell how meaningful the description of the data and the conclusions drawn about it are. This is where probability comes in. Scientists trying to decide if a treatment is effective want to know whether the patients got better because of the medicine or if it was just some freaky coincidence that everyone felt better. Measuring how likely events are (and whether things happened because of chance or because something is really happening) is where probability and statistics meet.

You might have been hoping that by learning probability that you'll be able to beat the house at games of chance. It's not going to happen any time in the near future. It requires a lot more than the basics of probability described in this chapter to make big money in Vegas. For one, the games are always stacked against the player, so the best way to make money in a casino is to own the casino. That's how you use probability to get rich.

But those are grand and illustrious goals. In this chapter we settle for some far more modest attempts to explain probability. With these more approachable goals, you'll be able to master simple games of chance—and to lay the foundation for more complicated calculations that await you in the future.

# Independent Events

We say that two events are independent if they don't have any influence on each other. Let's look at an example where we're flipping coins. I have a coin and you have a coin and we are flipping them. Flip, flip, flip. My coin does not care whether your coin comes up heads or tails. It just can't be bothered to worry about it, so it does its own thing. Heads, tails, whatever. My coin and your coin don't influence each other, so the flipping of my coin and the flipping of your coin are independent events.

It's easy to accept the fact that our flipping coins are independent. But flipping my coin again and again and again is also a sequence of independent events. Sure, it's my coin that's getting all the attention, but each flip is independent from the next. Heads or tails? It doesn't influence what my coin does next. It doesn't care if the previous flip was heads. It doesn't care if the previous flip was tails. It will keep coming up randomly, heads or tails.

Most people can accept this until they are faced with a situation like the following: my coin (and at this point I should point out that it is a fair coin, completely unbiased) has been tossed repeatedly and someone has been writing down what comes up. Weirdly, my coin has come up tails 20 times in a row! I'm about to flip the coin once more: which is more likely, heads or tails?

At this point someone is going to think that it should come up heads because heads is "due." Someone else is going to vote for tails because it's on a streak. Neither of them is right. My coin is just as likely to come up heads as it is tails. While this may be hard for some people to accept, you need to come to terms with this if you're going to work with probability and independent events.

## Coins

Probability problems almost always start with coins because everyone is familiar with coin flipping as a way to make random choices. Can't decide between vanilla or chocolate? Flip a coin. Want to decide which team goes first? Flip a coin. The coin flip is the standard way to make decisions when you have two choices.

The question becomes how do we quantify this in terms of probability? You're probably already familiar with terms like 50-50 that describe something about a coin toss, but you might not know exactly what is being talked about.

The way to mathematically quantify probability is to assign every possible event a number from zero to one. The number represents how likely that event is to happen. Things that aren't going to happen have probability zero. Things that will certainly happen have probability one. Everything else is in between. If the probability is close to zero, then the event is unlikely. If the probability is close to one, then the event is very likely. And if the probability is close to $\frac{1}{2}$, then the chance that the event will happen is close to the probability that the event won't happen. It could go either way.

Have you seen that TV show with the nerdy guys and the hot women? What is the chance that any of those women would be seen anywhere near those guys if it weren't part of the TV show? Probability zero. What about the chance that my mother will come visit my house and will complain that it isn't clean according to her demanding standards? Probability one. The chance that at least one of the jokes in this book will make you laugh? Somewhere in between. (We hope it's closer to one than zero.)

Your task in all of these probability problems is to quantify where the probability lies. A few things to remember: the answer has to be between zero and one, so if you get a negative number or something bigger than one, then you have made a mistake, and you need to go back and check your work. Also, because you're going to be working with numbers between zero and one, this means that you're going to be working with fractions and decimals. If you're not up to speed on those, then you need to detour back to Chapter 4. If you stick with probability without mastering fractions, you're wasting your time. I'm just sayin'.

In order to quantify probability, you need to learn about some fancy, mathematical ways of counting. Just about every probability problem you'll see at this level will boil down to counting two things. One of those things has the fancy name "sample space." It has nothing to do with the samples of food that they give out at the warehouse bulk shopping place on the weekends. It has nothing to do with the cross-stitch samplers that your great-aunt made, either. I have no idea why the word "sample" is in the term at all. It also has nothing to do with space. No outer space. No space invaders. No empty space. So, yeah, the sample space is a very badly named mathematical idea.

So now that you know the sample space isn't a free sample and it isn't from outer space, what is it? It is simply the collection of every single possible way that your event could happen. And all of these ways have to be equally likely.

A sample space with only one event is really boring. It's like a rigged election in a country where there is only one candidate who always runs unopposed. Sure, you could check to see what the outcome is, but you already know it and there isn't any uncertainty involved. But, while simple, a sample space with two equally likely events is more interesting.

So how do the coins come into this? And how do you find a sample space? And what is an event anyway? Back to the coins: if you have a fair coin, then when you flip it, there are two choices: heads or tails. An event is just a fancy way to say "something that happens." If you're looking at the problem of coin flipping as a whole, the overall event is (wait for it) coin flipping. The events that are the possible outcomes are "heads" and "tails." And your sample space consists of the two options: heads and tails. All that jargon just to say that there are two ways that a coin flip could come out: heads or tails.

So now that you know what the sample space is for coin flipping, you are ready to start thinking about how to calculate probabilities. Remember that probability is expressed as a number from zero to one. The denominator is the size of the sample space (in the case of the coin flip, this means that the denominator is 2). The numerator is the number of ways that the desired event could happen.

**Example:** You flip one coin. What is the probability that you get heads?

**Solution:** There are two elements of the sample space (heads and tails) and there is one way to get heads. Thus, the probability is $\frac{1}{2}$.

**Example:** You flip one coin. What is the probability that you get a lima bean?

**Solution:** A lima bean? That isn't in the sample space at all, so there is no chance of it happening. Therefore, the probability is $\frac{0}{2} = 0$.

So where does this notion of independent events come in? One event by itself is about as independent as you can be. But what happens when you flip the coin twice? A lot of times the language used to describe these situations is a little bit vague, so I want to be very, very clear what I mean here. When I say that you flip a coin twice, what I mean is that you flip the coin, record whether you saw heads or tails, flip it again, and then record what you saw again.

Flipping a coin twice represents independent events. The first flip doesn't talk to the second flip, and the second flip can't interfere with the first. They have no influence on each other, so they are independent events.

**Example:** You flip a coin twice. What's the probability that you get heads both times?

**Solution:** It all comes down to counting the sample space. If you flip a coin twice, what are all the possible things that could happen? You could get HH, HT, TH, or TT. That's it. So there are four items in your sample space. This means that when you set up your fraction to describe the probability, your denominator is going to be 4. What about the numerator? You have to count how many ways you can get heads both times. Well, that's kind of silly. By looking at the list (or using your common sense), you see that there is only one way to get heads both times: HH. Therefore, the probability is ¼.

> **Practice Makes Perfect**
>
> Problem 1: You flip a coin in each hand. The first comes up heads. What is the probability that the second one comes up tails?
>
> Problem 2: You flip a coin four times. What is the probability that you will get tails exactly three times?

# Dice

After flipping coins, the next classic example of probability that you see is dice. They're a little bit more complicated than coins, but not much. If you don't think you're ready to advance from coins to something more complicated, just think of dice as six-sided coins.

Once again, we're looking at equally likely events. Before, with coins, there were two events: heads and tails. Now there are six possible events: you can roll a one, two, three, four, five, or six. And each of these is equally likely. Just because you like one of these numbers better than the others doesn't make it any more or any less likely to come up. Just because a number is especially small or especially large doesn't make it any more or any less likely to come up. There's nothing special about a number that seems more average or typical than the others. When you roll one die, any of the six numbers will be equally likely to show up.

 **Amy's Answers**

The singular of dice is die.

If you don't believe me, here's what you do: go and get a die and roll it a thousand times and keep track of what numbers come up. At the end, tally the results. You'll see that each of the numbers from one to six comes up about the same number of times. They won't come up exactly the same number of times, but it will be relatively close.

**Example:** You roll a die. What is the probability that you get a four?

**Solution:** There are six elements of the sample space: one, two, three, four, five, and six. This means that the denominator of your fraction is going to be 6. Of those six events, only one of them is four, so the numerator will be 1. Therefore, the probability is $\frac{1}{6}$.

**Example:** You roll a die. What is the probability that you roll an even number?

**Solution:** There are still the same six elements of the sample space, so the denominator of your probability fraction is still going to be 6. What has changed in this problem is that there is more than one way for the desired event to happen. You can get an even number by rolling a two, a four, or a six, so there are three ways to do it. Therefore, the numerator in your probability fraction is going to be 3. The probability is $\frac{3}{6} = \frac{1}{2}$.

**Example:** You roll a die. What is the probability that you roll a five or a six?

**Solution:** We still have six elements in the sample space. Seems to be a theme in these problems, no? In this example, there are two events that we're looking for: rolling a five or rolling a six. Therefore the probability is $\frac{2}{6} = \frac{1}{3}$.

**Example:** You roll a die. What is the probability that you roll an 8?

**Solution:** This is one of those trick questions. It's impossible to roll an 8. Therefore, the probability is 0.

So what happens if you roll two dice? To make it simpler to talk about them, let's pretend that they're different colors, one is yellow and the other is purple. Just like the two coins, the two dice are going to be independent events because they don't influence each other. They just don't care. If you get a three on the yellow die, the purple die could come up with any of the numbers from one to six, and none will be more or less likely than any of the others.

In order to answer questions about the probability of events involving two dice, it's really helpful to make a table that shows all of the equally likely events in the sample space. You'll want to go about doing this in an organized way because there are a lot of things that could happen. In fact, there are 36 things that could happen. It's no coincidence that $36 = 6^2$.

The following table shows all of the possible results of rolling two dice. The first number shows the result of the yellow die, and the second number shows the result of the purple one. It's pretty clear that having a 1 come up on the yellow and a 5 on the purple is going to be different than a 1 on the purple and a 5 on the yellow. However, if they both come up 3, then you can't tell which is which, so there is only one way it can happen.

| | | | | | |
|---|---|---|---|---|---|
| 1,1 | 1,2 | 1,3 | 1,4 | 1,5 | 1,6 |
| 2,1 | 2,2 | 2,3 | 2,4 | 2,5 | 2,6 |
| 3,1 | 3,2 | 3,3 | 3,4 | 3,5 | 3,6 |
| 4,1 | 4,2 | 4,3 | 4,4 | 4,5 | 4,6 |
| 5,1 | 5,2 | 5,3 | 5,4 | 5,5 | 5,6 |
| 6,1 | 6,2 | 6,3 | 6,4 | 6,5 | 6,6 |

This chart shows all of the possibilities in the sample space. This means that in all of the problems we do involving two dice, the denominator of the probability fraction is going to be 36.

**Example:** You roll two dice. What is the probability that the yellow die comes up 3 and the purple one comes up 6?

**Solution:** In all of these problems, the denominator is going to be 36. Looking on the chart above, you can see that there is only one way for the yellow to be 3 and the purple to be 6. Therefore, the probability is $1/36$.

> **Practice Makes Perfect**
>
> Problem 3: You roll two dice. What is the probability that you get 5 on the purple one, but not the yellow one?

**Example:** You roll two dice. What is the probability that one of them comes up 3 and the other one comes up 6?

**Solution:** This question is a slight variation on the previous one. The key difference is that the previous question specified what number came up on which color die. This question doesn't care. Just like before, your denominator is going to be 36. In this case, however, there are two ways for your event to happen. You can get a 3 on the yellow and a 6 on the purple. Or you can get a 3 on the purple and a 6 on the yellow. Thus, there are two ways for your event to happen. This means that your probability fraction is going to be $2/36 = 1/18$.

**Example:** You roll two dice. What is the probability that the sum of the numbers showing is 8?

**Solution:** Just like all of these problems, there are the 36 possibilities in the chart, so you know that the denominator of your probability fraction is going to be 36. To find the numerator, you need to count how many options in the chart have the two numbers adding up to 8. You should see that 2 + 6 = 8, 3 + 5 = 8, 4 + 4 = 8, 5 + 3 = 8, and 6 + 2 = 8, so there are five ways to do it. Thus, your answer is $5/36$.

# Cards

Way back a long time ago before there was the Internet and when television only had a few channels, families used to get together in the evenings to play cards. Other wholesome activities, like ice-skating and board games, were also popular, but what I really want to talk about is card games. Back then everyone knew how many cards were in a deck and what the different cards were like. These days people don't play cards quite so much. Some families are too busy. Other families have religious reasons for not playing cards. There are probably a bunch of other reasons, too. Nevertheless, this lack of card playing means that I need to start this section by describing what is in a standard deck of playing cards. (Don't laugh! There are people who don't know this!)

A deck of cards consist of 52 cards that are divided into four suits: hearts, diamonds, clubs, and spades. Hearts and diamonds are red; clubs and spades are black. There are 13 cards in each suit. (This should make sense because $4 \times 13 = 52$.) The cards in each suit are numbered from two to ten, and the remaining four cards are the ace (often thought of as "one"), the jack, the queen, and the king. The jack, queen, and king are called "face cards."

What makes the card examples more difficult than the dice examples is that now we have 52 objects to deal with, and they have all sorts of properties: color, suit, and number.

Fortunately, the methods you've been using are still going to work. You just need to identify how many elements are in the sample space and then decide how many of the elements in the sample space represent successes.

If you're given a problem about the entire deck of cards and you're picking one of them at random, then your sample space consists of all the cards. Because there are 52 of them, your sample space will have 52 members, so the denominator of your probability fraction will be 52.

**Example:** You pick a card at random from a shuffled deck. What's the probability that you pick the queen of spades?

**Solution:** There are 52 cards in the deck, so the denominator is 52. Only one of them is the queen of spades, so the probability is $1/52$.

**Example:** You pick a card at random from a shuffled deck. What's the probability that you pick a seven?

**Solution:** Once again, the denominator will be 52. How many sevens are in the deck? There are four of them: the seven of hearts, the seven of spades, the seven of clubs, and the seven of diamonds. Therefore, the probability is $^4/_{52} = {}^1/_{13}$.

**Example:** You pick a card at random from a shuffled deck. What's the probability that it's a heart?

**Solution:** Since 13 of the 52 cards are hearts, the probability is $^{13}/_{52}$.

## Chances of Not Happening

One remaining useful fact about probability is how to find the probability that something doesn't happen if you know the probability that it does happen. It works by a simple formula. If the probability that an event will happen is $p$, then the probability that it won't happen is $1 - p$.

**Example:** You pick a card at random from a deck. What is the probability that it is not the queen of spades?

**Solution:** There are 52 cards in the deck, and only one of them is the queen of spades. So the probability that you pick the queen of spades is $^1/_{52}$. However, that's not what the question was asking. To find the probability that the card is *not* the queen of spades, you need to subtract $1 - {}^1/_{52} = {}^{51}/_{52}$.

## The Least You Need to Know

- Probability is a measure of how likely it is that an event occurs, will occur, or occurred.

- Rolls of a die and flips of a coin are independent events. The previous roll of a die tells you absolutely nothing about what the next roll will be. The same is true with flipping coins.

- To find the probability of an event, for example rolling a seven on two six-sided dice, look at every possible outcome, and divide the number of favorable outcomes by the total number of all possible outcomes.

- The probability of drawing a specific card from a deck of cards is $\frac{1}{52}$.

Appendix **A**

# Glossary

**absolute value**   How far away a number is from zero.

**abstraction**   Taking a concept and making it have nothing whatsoever to do with reality.

**altitude**   The height of a triangle. The altitude is perpendicular to the base of the triangle and connects the base to the opposite corner.

**area**   How much space is inside the boundaries of a shape.

**array**   A rectangular arrangement of numbers or objects.

**Associative Law**   States that when adding or multiplying, the way you group the numbers doesn't matter. For example, when multiplying $3 \times 2 \times 5$, you can do $(3 \times 2) \times 5 = 6 \times 5 = 30$ or you can do $3 \times (2 \times 5) = 3 \times 10 = 30$.

**balance**   Amount of money in an account or amount owed on a loan. Often seen in financial word problems.

**base**   The bottom number in an expression with an exponent. In geometry, the bottom of a triangle or another geometric figure.

**binomial**   A polynomial with two terms.

**Cartesian coordinates**   A way to describe locations in the plane by two numbers. Invented by Descartes.

**circumference**   The perimeter of a circle.

**Commutative Law**   A law that states that the order of the numbers doesn't matter when multiplying the numbers. For example, multiplying 3 and 4 is the same as multiplying 4 and 3. The same law applies to addition. Order doesn't matter.

**complex fraction**   A fraction on top of a fraction.

**composite number**   Any positive integer that's not a prime number.

**common multiple**   A number that your two given numbers both divide into evenly.

**conversion factor**   Used for converting one type of unit to another, such as miles to kilometers.

**coordinate axis**   A directed line that's used to describe the locations of objects in the plane. The equivalent of the number line when you're in a plane.

**coordinate plane**   The Cartesian plane.

**cube root**   What you need to multiply by itself a total of three times to get the requested number. For example, the cube root of 64 is 4 because $64 = 4 \times 4 \times 4$.

**decimal**   A way of representing numbers that may be between whole numbers. Decimal numbers are written with the digits 0 through 9, at most one decimal point, and no other mathematical symbols.

**decimal places**   In a decimal number, the digits after the decimal point.

**denominator**   The bottom part of a fraction.

**density**   The amount of mass per unit volume.

**diameter**   The distance across the circle at the widest point.

**difference**   The result of a subtraction.

**distance formula**   Finds the distance between two points.

**Distributive Law**   A rule that explains how to multiply by things that are added together. It says that $a(b + c) = ab + ac$ and $(a + b)c = ac + bc$.

**divisible**   Translates to "can be divided by." One number is divisible by another number if the second number goes evenly into the first without leaving a remainder.

**equations**   Mathematical sentences in which two clumps of mathematical symbols are joined with an equals sign.

**equilateral triangle**   A triangle where all the sides have the same length.

**equivalent fractions**   Fractions that look different but have the same value. For example, $\frac{1}{2}$ is the same as $\frac{2}{4}$.

**exponent** The small number up and to the right of a number that tells you how many times to multiply it by itself. Also called the power. For example, seven to the fourth power is written $7^4$, which is the same as $7 \times 7 \times 7 \times 7$.

**expression** A bunch of letters and numbers joined together with mathematical operations.

**factor** Something that divides evenly into a number or expression.

**factor tree** A fun picture used to keep track of how a number is factored.

**factoring** Breaking a number down into its factors.

**FOIL** An acronym used to remember how to multiply together two binomials: first, outside, inside, last.

**fractions** Numbers that are represented as one number (called the numerator) over another (called the denominator). Fractions are used to show how a whole has been divided into parts.

**glide reflection** To flip and then slide something. A pattern with a structure like footprints, where the mirror image of the image is offset from the original.

**greatest common factor (GCF)** The biggest factor two numbers have in common.

**gross profit** Selling price minus cost.

**height** How tall a shape is, measured from the base.

**hypotenuse** The longest side of a right triangle. Located directly opposite from the right angle.

**identity element** A number that leaves things unchanged when you add it or multiply by it. When adding, it's zero. When multiplying, it's one.

**imperfect square** A number that cannot be written as a whole number times itself.

**improper fractions** A fraction with a bigger number above the fraction bar and a smaller number on the bottom.

**integers** All of the natural numbers, their negatives, and zero.

**interest** Money paid to holders of savings accounts or by debtors to lenders.

**interior angles** The angles on the inside of a shape.

**inverse operations** The mathematical version of undo. One example is adding a number and subtracting off the same number.

**irrational number**   A number that can never be written as a fraction with both the numerator and the denominator being integers.

**irregular polygon**   Not all the sides are the same length and/or not all the angles are equal.

**isosceles triangle**   A triangle where two sides have the same length.

**kite**   A quadrilateral shaped like a kite. It has two pairs of sides that are the same length. Each side must be next to a side that is the same length as itself.

**least common multiple (LCM)**   The smallest multiple two numbers have in common.

**lowest common denominator**   The least common multiple of the denominators of the fractions. Necessary to add or subtract fractions.

**lowest terms**   A fraction where the numerator and denominator of the fraction have no common factors.

**markup**   Gross profit above cost as a percentage of the cost.

**mass**   How much matter there is in an object.

**midpoint formula**   A formula that finds the point halfway between two points.

**mirror symmetry**   What you get when you look in a … mirror. A pattern in which one side matches with the other side when the pattern is folded in half.

**mixed number**   A number written as a whole number together with a fraction. There is an invisible plus sign between the whole number and the fraction.

**monomial**   A mathematical object in which variables are raised to whole number powers and multiplied by numbers called coefficients.

**multiple**   Obtained by multiplying by a number. If you multiply something by seven, the result is a multiple of seven.

**natural numbers**   The first thing that comes to mind when you think numbers. They must be bigger than zero and not require decimals or fractions to write them. All the numbers you can get by adding 1 to itself repeatedly. These are sometimes called the counting numbers: 1, 2, 3 ….

**negative**   Not positive. Less than zero.

**number line**   An imaginary line that keeps track of numbers and their relative sizes. Zero is in the middle, negatives to the left, positives to the right.

**numerator**   The top part of a fraction.

**order of operation**   The mandatory order in which you perform math operations.

**origin**   The point at which the coordinate axes meet.

**parallelogram**   A quadrilateral with two sets of parallel sides.

**percent**   A fraction with a denominator of 100.

**perimeter**   The distance around the whole outside of a shape.

**pi**   A fun number. Used in circles. The ratio of a circle's circumference to its diameter.

**polygon**   A two-dimensional shape made out of straight edges. Has a distinct inside and outside.

**polynomial**   A sum of monomials.

**power**   *See* exponent.

**prime decomposition**   *See* prime factorization.

**prime factorization**   A way to break a number down to find all of its prime factors.

**prime number**   A number divisible only by 1 and itself.

**principal**   The original amount of money invested or borrowed. Often seen in financial word problems.

**profit margin**   The profit above cost as a percentage of the selling price.

**proportion**   Quantities that have a multiplicative description of how the part relates to the whole.

**Pythagorean Theorem**   One of the most famous theorems about right triangles. The sum of the squares of the lengths of the legs equals the square of the length of the hypotenuse, usually expressed as $a^2 + b^2 = c^2$.

**quadrilateral**   A polygon with four sides.

**quotient**   The answer to a division problem.

**radical**   The square root symbol.

**radius**   Half of the diameter. The distance from any point on the circle to the center.

**ratio**   A multiplicative relationship between two quantities, often expressed as a quotient.

**rational number**   Any number that can be written as the quotient of two integers.

**real numbers**   All the numbers that are possible to write with our decimal system.

**reciprocal**   The reciprocal is the fraction you get when you interchange the numerator and denominator of a fraction.

**reflection**   *See* mirror symmetry.

**regular polygon**   A polygon in which all the sides are the same length and all of the angles have the same measure.

**remainder**   When dividing gets messy, this is the number left over.

**rhombus**   A quadrilateral where all the sides have the same length. Yes, a square is one.

**right triangle**   A triangle in which one of the angles is a right angle, 90 degrees.

**rotation**   When you turn something. A type of symmetry where the pattern will line up again after being rotated.

**scalene triangle**   A triangle where all the sides have different lengths.

**significant digit**   In a number representing a measurement, a significant digit is a digit that's value isn't prone to error because of inaccuracies of your measurement.

**simplest terms**   Another name for lowest terms.

**skip-counting**   Counting by skipping the same number of numbers.

**solution**   A value that a variable can take on to make an equation true.

**square root**   What you need to multiply by itself to get the requested number.

**squaring a number**   Multiplying a number by itself.

**surface area**   How much area is on the surface of a solid in three-dimensional space. If you were going to wallpaper a shape, the amount of paper you'd need.

**symmetry**   One of the ways the universe is balanced. A way that a pattern can be repeated so that its parts match up.

**terminating decimals**   Decimals that have a finite number of decimal places.

**translation**   Sliding something around in the plane.

**trapezium**   A fancy name for a trapezoid.

**trapezoid**   A quadrilateral with one pair of parallel sides.

**unit**   The standard by which things are measured, such as feet or kilograms.

**unit multiplier**   A ratio by which you multiply when converting from one type of measurement to another.

**variable**   A letter that stands for a number.

**vector**   A mathematical quantity that has both a length and a direction.

**volume**   The three-dimensional size. The amount of space an object takes up.

**whole numbers**   The natural numbers together with zero.

**x-axis**   The horizontal axis on a graph.

**y-axis**   The vertical axis on a graph.

# Solutions to Practical Practice Problems

## Chapter 1

### Problem 1:

17 and 94,439,238 are both natural numbers. 17 can be reached by adding 1 seventeen times. 94,439,238 can be reached by adding 1 ninety-four million, four hundred thirty-nine … you get the picture. –845 and 16.4 are not natural numbers because you cannot add 1 any number of times to reach either number.

### Problem 2:

17 is a prime number, so its only factors are 1 and 17. To prove this, try dividing 17 by 2, 3, 4, etc.

### Problem 3:

The factors of 48 are 2, 3, 4, 6, 8, 12, 16 and 24. Numbers such as 5, 7, 9, and 11 are not factors of 48 because they do not divide evenly into 48.

### Problem 4:

Take two factors of 52: 4 and 13. 13 is already a prime number. 4 can be factored to $2^2$ and 2 is prime. Therefore, the prime factorization of 52 is $2 \times 2 \times 13$. The same answer can be found if you start with 2 and 26 as the two factors of 52.

**Problem 5:**

The number 38 only has two factors other than 1 and itself: 2 and 19. The prime factorization of 38 is $2 \times 19$ since both 2 and 19 are prime numbers.

**Problem 6:**

Start by making a list of all the factors:

40: 1, 2, 4, 5, 8, 10, 20, 40

60: 1, 2, 3, 4, 5, 6, 10, 12, 15, 20, 30, 60

Looking at the list you can see that 20 is the largest number on both lists.

You can also use the prime factorization of the numbers:

40: $2 \times 2 \times 2 \times 5$

60: $2 \times 2 \times 3 \times 5$

The common prime factors are: $2 \times 2 \times 5 = 20$.

**Problem 7:**

First find the prime factorization of each number:

52: $2 \times 2 \times 13$

60: $2 \times 2 \times 3 \times 5$

52 and 60 both have two 2s, so the greatest common factor is $2 \times 2 = 4$.

**Problem 8:**

First, list the multiples of each number:

15: 15, 30, 45, 60, 75, 90, etc.

20: 20, 40, 60, 80, 100, etc.

60 is the first number that appears on both lists, so 60 is the least common multiple.

**Problem 9:**

List the multiples of each number:

7: 7, 14, 21, 28, 35, 42, 49, 56, 63, 70, 77, 84, etc.

12: 12, 24, 36, 48, 60, 72, 84, 96, 108, 120, etc.

The smallest number on both lists is 84. It is important to note that $84 = 7 \times 12$. When two numbers do not have any factors in common, the least common multiple is the product of the two numbers.

**Problem 10:**

List the multiples of each number:

10: 10, 20, 30, etc.

16: 16, 32, 48, 64, 80, 96, etc.

The first time that a multiple of 16 is a multiple of 10 is when 16 is multiplied by 5. So the least common multiple of 10 and 16 is 80.

**Problem 11:**

12 is a positive number, so leave it alone. The absolute value is 12!

**Problem 12:**

$|-53| = 53$. $-53$ is 53 steps away from 0.

**Problem 13:**

The distance between $-0$ and 0 is still 0. There isn't really even a number $-0$, in case you were a little confused.

**Problem 14:**

Both numbers are positive, so there is no special trick to remember here. $13 + 5 = 18$

**Problem 15:**

13 is greater than 5, so when you subtract 5, you still get a positive number. $13 - 5 = 8$.

**Problem 16:**

This is the same as $-(13-5) = -7$. $-13$ is negative and 5 is positive, but 5 is less than 13, so you should expect the answer to be negative.

**Problem 17:**

This is like adding $(-13) + (-5)$. You can think of this as the negative of adding the positives, or: $-(13 + 5) = -18$.

**Problem 18:**

Subtracting a negative number is the same as adding a positive number. $-13 - (-5) = -13 + 5$. Looking back to Problem 17, we know that $-13 + 5 = -7$.

**Problem 19:**

Subtracting a negative number is the same as adding a positive number, so the equation is equal to 13 + 5 = 18.

**Problem 20:**

Adding a negative is the same as subtracting a positive. Because 5 is less than 13, we can subtract in a normal way: 13 – 5 = 8.

**Problem 21:**

In this case, the expression is equivalent to: – (13 + 5) = –18. When you add a negative, a number should get smaller, so it makes sense that adding –5 to –13 equals –18, which is less than –13.

**Problem 22:**

11 and 8 are both positive, so their product is positive. 11 x×8 = 88.

**Problem 23:**

Both numbers are negative, so their product is positive. –13 × (–4) = 4 × 13 = 52.

**Problem 24:**

–9 is negative, but 2 is positive. Multiply them as if they were both positive: 9 × 2 = 18. Now take the negative, as in the previous problem: –9 × 2 = –18.

# Chapter 2

**Problem 1:**

(a) –73 is an integer and a real number. –73 doesn't have any numbers after a decimal point, so it is an integer. It is not a whole number because it is less than zero. It is a real number because it can be written with the number 0-9. It is *not* a natural number because it is negative.

(b) 0 is an integer, a whole number, and a real number. Because 0 doesn't have any numbers after a decimal point, it is an integer. It is a whole number because it is 0 or a natural number. It is not a natural number because it is less than 1.

(c) –0.001 is a real number. Because all whole numbers, integers, and natural numbers have no numbers after a decimal point, it cannot be any of these.

(d) 43 is an integer, a whole number, a natural number, and a real number. Because 43 doesn't have any numbers after a decimal point, it is an integer. It is a natural number because it can be written as the sum of 1's. It is a whole number because it is 43 or a natural number.

**Problem 2:**

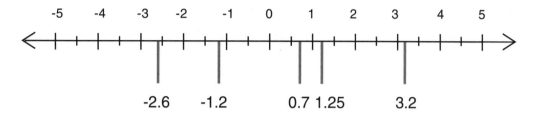

**Problem 3:**

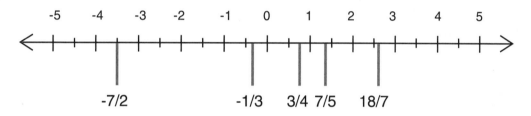

# Chapter 3

**Problem 1:**

(a) Evaluate the inner-most expression, which is (2 – 4) = –2 in this case. Add to 7: 7 + (–2) = 5. Now multiply this number by 8. 5 × 8 = 40. Finally, add 4. 40 + 4 = 44.

(b) Watch carefully on this one. The inner-most parentheses is now (7 + 2) = 9. Subtract 4 from 9 to get 5. Multiply 5 and 8 to get 40 and add 4 to get 44. Note that this is the same as in problem (a). This makes sense because addition is commutative and in both cases, you add (–4), so the solution of [7 + (2 – 4)] = [(7 + 2) – 4].

(c) This problem is a little different, so don't get fooled into thinking it will be the same as the first two problems. First, add (7 + 2) = 9. Then, multiply by 8. Do not subtract 4 yet! 8 × 9 = 72. Now you can subtract 4. 72 – 4 = 68. Finally, add 4 to get a final answer of 72. Note that adding 4 "cancels" out the subtraction of 4.

**Problem 2:**

(a) Find the first layers and begin unpeeling your onion. Evaluate (2 × (14 – 3)) by subtracting in the inner parentheses and then multiplying. 2 ×(14 – 3) = 2 × 9 = 18.

Subtract this from 16. 16 − 18 = −2. Now work on the other set of parentheses. 8 × 3 ÷ 2 = 12 whether you multiply or divide first. Now subtract 2. 12 − 2 = 10. Add (−3 × 6) = −18 to this. 10 + (−18) = −8. Evaluate (−13 − 8)= −21. Multiply this by −8. −21 × −8 = 168. Now look back and find your answer from the first group of parentheses (−2). Now you can add this to 168. 168 + (−2) = 166.

(b) Evaluate expressions in the following order:

Left side:

1. (2 × 14 − 3) = 28 − 3 = 25

2. 16 − 25 = −9

Right side:

3. (8 × 3 ÷ 2) − 2 = 12 − 2 = 10

4. (−3 × 6) = −18

5. ((8 × 3 ÷ 2) − 2) + (−3 × 6) = 10 + (−18) = −8

6. 8 × (−8) = −64

7. −13 − (−64) = 51

8. (−9) + 51 = 42

Finally, the final answer is 42!

(c) In this problem, evaluate all of the expressions in parentheses and then remember to multiply before adding and subtracting. The following steps start from the left side and go to the right:

(14 − 3) = 11

2 × (14 − 3) = 2 × 11 = 22

16 − 2 × (14 − 3) = 16 − 22 = (−6)

(−3 × 6) = −18

2 + (−3 × 6) = 2 + (−18) = −16

(8 × 3 ÷ 2) = 12

(8 × 3 ÷ 2) − (2 + (−3 × 6)) = 12 − (−16) = 28

(−13 − 8) = −21

$(-21) \times (28) = -588$

$(-6) + (-588) = -594$

And the final answer is ... (drum roll) ... –594!

(d) $16 - (2 \times (14 - (3 + (-13 - (8 \times (18 \times 3) \div (2 - (2 + (-3 \times 6)))))))))$

Make sure to find the inner-most parentheses! Start at the right and work toward the left.

$(-3 \times 6) = -18$

$2 - (2 + (-18)) = 2 - (-16) = 18$

$(18 \times 3) = 45$

$8 \times (45) = 360$

$(360) \div (18) = 20$

$3 + ((-13) - 20) = 3 + (-33) = -30$

$14 - (-30) = 44$

$2 \times (44) = 88$

$16 - (88) = -72$

And the final answer is (–72).

## Problem 3:

Remember that addition is commutative, so the answer is the same whichever way you add it. $8 + 4 = 4 + 8 = 12$.

Subtraction is not commutative, so the answer will be different. $10 - 7 = 3$. $7 - 10 = -3$. Note that the answers are opposites.

Multiplication is commutative, so $4 \times 6 = 6 \times 4 = 24$.

## Problem 4:

The Associative Law holds true for addition, so $(7 + 8) + 5 = 7 + (8 + 5) = 20$

Subtraction does not work with the Associative Law. Check to see that they are different: $(24 - 5) - 12 = (21) - 12 = 9$ and $24 - (5 - 12) = 24 - (-7) = 31$.

Multiplication works under the Associative Law. Check: $(2 \times 4) \times 5 = 8 \times 5 = 40$ and $2 \times (4 \times 5) = 2 \times (20) = 40$.

Make sure to follow the order of operations! $(14 + 6) \times 7 = (20) \times 7 = 140$ and $14 + (6 \times 7) = 14 + (42) = 56$. The two are quite different. This shows how important it is to follow the order of operations!

**Problem 5:**

Evaluate the part inside the parentheses.

$6 \times (17) = 102$.

**Problem 6:**

Evaluate the part inside the parentheses first.

$11 \times (4) = 44$.

**Problem 7:**

Evaluate the parentheses first.

$(5) \times 5 - 25$.

**Problem 8:**

Multiply $(-6)$ by 4 and then by 5. $-6 \times (4 + 5) = (-6)(4) + (-6)(5) = -24 + (-30) = -54$.

The minus sign is the same as multiplying by $(-1)$. Therefore, $-(-11 - 7) = (-1)(-11) + (-1)(-7) = 11 + 7 = 18$.

Whether the $(-9)$ is to the left or the right, you can still distributive it across the numbers inside the parentheses. $(6)(-9) - (5)(-9) = -54 - (-45) = -9$.

Distribute the minus sign across the expression and then distribute the $(-11)$:

$-(8 - 12) = (-1)(8) + (-1)(-12) = (-8 + 12)$.

Distribute the $(-11)$ now:

$(-11)(-8 + 12) = (-11)(-8) + (-11)(12) = 88 - 132 = -44$.

**Problem 9:**

In this problem, it doesn't matter whether you add or subtract first. However, you can see that adding 45 and then subtracting 45 is the same is adding 0. $45,353 + (0) = 45,353$.

**Problem 10:**

The order of operations states that you multiply and divide left to right. $672{,}834 \times 253 = 170{,}227{,}002$. Now divide. $170{,}227{,}002 \div 253 = 672{,}834$.

**Problem 11:**

Evaluate the expressions inside the parentheses first. $(234 - 37) = 197$ and $(37 - 234) = -197$. You can see that $197 + (-197) = 0$, so the final answer is $9{,}834 + (0) = 9{,}834$.

**Problem 12:**

Evaluate the parentheses, then multiply and divide. $(56 - 21) = 35$. $(-23) \times (35) - 23 = -35$. Now add and subtract the right side. $15 + 56 - 21 = 50$. Add this to $(-35)$. $(-35) + 50 = 15$.

# Chapter 4

**Problem 1:**

(a) The numerator is greater than the denominator, so this is an improper fraction. To change it to a mixed number, divide and find the remainder. 5 goes into 23 four times with three left over. So $^{23}\!/_5 = 4^3\!/_5$.

(b) Because there is an integer and a fraction, this is a mixed number. Multiply the integer by the denominator of the fraction and add it to the numerator. $5 \times 7 + 5 = 40$. The fraction is equal to $-^{40}\!/_7$. In words, this is negative forty sevenths (not negative forty-*seven!*).

**Problem 2:**

(a) To cross-multiply, multiply the numerator of each with the denominator of the other fraction. $3 \times 14 = 42$. $12 \times 4 = 48$. Because the two are unequal, the fractions are not equivalent.

(b) $5 \times 14 = 70$. $7 \times -10 \doteq 70$. The two are not equal, so the fractions are not equivalent.

(c) $0 \times 100 = 0$ and $2 \times 0 = 0$. Because both fractions are equal to 0, the fractions are equivalent.

**Problem 3:**

(a) To reduce, divide both the numerator and the denominator by 3. $6 \div 3 = 2$. $9 \div 3 = 3$. So $^6\!/_9 = {}^2\!/_3$. You cannot reduce any further.

(b) Divide the both by 4. $4 \div 4 = 1$. $8 \div 4 = 2$. Therefore, $^4\!/_8 = {}^1\!/_2$.

(c) Both the numerator and the denominator are divisible by 3. $15 \div 3 = 5$. $9 \div 3 = 3$. So $^{15}/_9 = ^5/_3$.

**Problem 4:**

(a) There are two digits to the right of the decimal point, so $0.47 = ^{47}/_{100}$. This fraction cannot be reduced.

(b) There are seven digits to the right of the decimal point, so $0.5684379 = $ $^{5,684,379}/_{10,000,000}$.

**Problem 5:**

(a) Because there is just one digit in the repeated pattern, $0.\overline{1}$ is equal to $^1/_9$.

(b) There are two digits in the pattern, so it is equal to $^{84}/_{99}$.

**Problem 6:**

Multiply both the tops and the bottoms. $9 \times 2 = 18$. $7 \times 5 = 35$. So the product of the two is $^{18}/_{35}$.

**Problem 7:**

$^2/_5 \div ^9/_7 = ^2/_5 \times ^7/_9 = ^{14}/_{45}$.

**Problem 8:**

(a) Because both fractions have the same denominator, you can simply add the numerators. The sum is $^{10}/_{13}$.

(b) Subtract the numerators and reduce. The difference is $-^4/_8 = -^1/_2$.

**Problem 9:**

Because the denominators are not the same, multiply $^1/_2$ by $^5/_5$ and $^1/_5$ by $^2/_2$. The new sum is $^5/_{10} + ^2/_{10} = ^7/_{10}$.

**Problem 10:**

You can multiply the first fraction by two and the denominators will be the same. $^4/_6 + ^7/_6 = ^{11}/_6$.

**Problem 11:**

You can multiply the first fraction by four and the second by three, and the denominators of both will be 24. $^5/_6 = ^{20}/_{24}$ and $^9/_8 = ^{27}/_{24}$. Now subtract the numerators. $^{20}/_{24} - ^{27}/_{24} = -^7/_{24}$.

# Chapter 5

**Problem 1:**

(a) To convert to decimals, move the decimal point two places to the left. 98% = 0.98.

(b) 0.005% = 0.00005

(c) 4.23% = 0.0423

(d) When a percent is greater than 100, then you multiply the quantity by a number greater than 1. 623% = 6.23

**Problem 2:**

To convert from decimal to percent, multiply by one hundred or move the decimal two places to the right.

(a) 0.37 = 37%

(b) 1.01 = 101%

(c) 3.3 = 330%

(d) 72 = 7200%

**Problem 3:**

Convert the fractions to percents by multiplying the top and bottom by whatever number makes the denominator 100, or convert to decimal and then percent.

(a) $3/4 = {}^{75}/_{100} = 0.75 = 75\%$

(b) $7/20 = {}^{35}/_{100} = 0.35 = 35\%$

(c) $29/25 = {}^{116}/_{100} = 1.16 = 116\%$

(d) $3^2/5 = {}^{340}/_{100} = 3.40 = 340\%$

**Problem 4:**

To convert to a fraction, put the percent over 100 and simplify.

(a) $25\% = {}^{25}/_{100} = 1/4$

(b) $9.8\% = {}^{9.8}/_{100} = {}^{98}/_{1,000} = {}^{49}/_{500}$

(c) $0.04\% = {}^{0.04}/_{100} = {}^{4}/_{1,000} = {}^{1}/_{2,500}$

(d) $355\% = {}^{355}/_{100} = {}^{71}/_{20}$

**Problem 5:**

Write the ratio of black pens to total number of pens as a fraction and then convert to a percent. $^{120}/_{125} = {}^{24}/_{25} = {}^{96}/_{100} = 96\%$. 96% of the pens on the desk are black.

**Problem 6:**

To find 25% of the books, multiply the total by 0.25. $160 \times 0.25 = 40$. The monster reads 40 books each morning!

**Problem 7:**

First, calculate 20% of the number of elephants. $5 \times 0.2 = 1$. Then, there is one more tiger than there are elephants, since there are 20% *more* tigers. There are 6 tigers in the zoo.

**Problem 8:**

For this problem, find the ratio of the world's population in 2006 to the population in 1956. $\dfrac{6.5 \text{ billion}}{2.8 \text{ billion}} = 2.32$. Convert this to a percent. $2.32 = 232\%$. Because the population increased, the percent is greater than 100.

**Problem 9:**

Calculating for the first year, Marvin has $2,500 + 2,500 \times 0.045 = 2612.5$. The second year, it continues to increase: $2612.5 + 2612.5 \times 0.045 = 2730.06$. Continue this process twice more to find that Marvin has $2852.92 after 4 years.

**Problem 10:**

Remember that 10% is like $^1/_{10}$ or like dividing by 10, moving the decimal point one place to the left. So 10% of 37.58 is 3.758. It might be hard to find 0.008 cents in your wallet, though, so you could just round up to $3.76.

**Problem 11:**

To calculate a 20% tip, first 10% of 78.65 and then double it. $^{78.65}/_{10} = 7.856$. Twenty percent is $7.856 \times 2 = 15.712$. $15.71 will be left on the table for the waiter.

**Problem 12:**

Gross profit: Assuming the retailer receives the gross profit as his commission, the gross profit is $0.15 \times 1,000 = 150$.

Profit margin: $^{150}/_{1,000} = 0.15 = 15\%$.

Markup: $^{150}/_{850} = 0.20 = 20\%$.

### Problem 13:

Find the difference in the two values and divide by the actual, accepted density. $^{0.2}/_1 =$ 0.2. Multiply by 100 to get a 20% error in measurement.

### Problem 14:

Again, find the difference in the two values, divide by the actual value (500 g), and multiply by 100 to express the error as a percent. $500 - 489 = 11$. $^{11}/_{500} = 0.022$. $0.022 \times 100 = 2.2\%$. Someone must have stolen 2.2% of 500 g of your peanuts!

# Chapter 6

### Problem 1:

Write the problem in the form $\dfrac{\text{Marvin's Weight}}{\text{Green's Weight}} \cdot \dfrac{160\text{lbs}}{8000\text{lbs}} = 0.02$.

### Problem 2:

Write the problem in the form $\dfrac{\text{number of ants}}{\text{number of people}} \cdot \dfrac{560 \text{ ants}}{80 \text{ people}} = 7$ ants per person.

### Problem 3:

The coke with the higher concentration of salt will taste saltier, not necessarily the coke with the most number of spoons of salt. Convert the volume of coke to the same unit, either both liters or both milliliters. Find the concentration of the first coke by dividing the amount of salt by the volume of coke. $\dfrac{3 \text{ spoons salt}}{500 \text{ mL}} = .006$ spoons per milliliter. For the second coke, convert liters to milliliters to find that the concentration is $\dfrac{10 \text{ spoons salt}}{2000 \text{ mL}} = .005$ spoons per milliliter. The concentration of the first coke is higher, so it will taste saltier. Either way, the coke probably won't taste great!

### Problem 4:

To make the soup taste the same, you want to make sure the ratio of the cooked chicken to the water is the same before and after adding water. Find the ratio before adding water. $\dfrac{4 \text{ cups cooked chicken}}{12 \text{ cups water}} = \dfrac{1}{3}$ cups of cooked chicken per cup of water. Then, add 18 cups of water to the amount already in the soup for a total of 30 cups of water. Some amount of cooked chicken divided by the 30 cups of water will equal $^1/_3$. $^{10}/_{30} = ^1/_3$. So there needs to be a total of 10 cups of cooked chicken, but there are already 4 cups of chicken in the soup. So add $10 - 4 = 6$ cups of chicken to the soup to make it taste delicious.

**Problem 5:**

First, find out the total cost of the bananas in Hong Kong dollars. 5 lbs × HK$3.30 = HK$16.50. Now multiply by the ratio $\dfrac{\text{US\$1}}{\text{HK\$7.79}}$. $(16.50) \times \frac{1}{7.79} = 2.12$. So the bananas will cost US$2.12.

**Problem 6:**

To find the weight of the sheep per person in Argentina, multiply the ratio by the weight of a sheep. $\frac{2}{1} \times 400 = 800$. This means that there are 800 lbs of sheep per person in Argentina. In New Zealand, however, there are $\frac{12}{1} \times 70 = 840$ lbs of cow per person. Therefore, New Zealand has a few more pounds of livestock per person than Argentina.

# Chapter 7

**Problem 1:**

Starting on the first square, the king would put 1 coin, 2 on the second, 4 on the third, and so forth. Turn all of these into powers of two by thinking of the first square as $2^0$, the second square as $2^1$, the third square as $2^2$ and so forth. When you get to the tenth square, you will have $2^9$, not $2^{10}$, since you started at $2^0$. Now evaluate $2^9$. You can use a calculator, or simply carefully calculate it by hand. $2^9 = 512$. And that's only the number of coins on the tenth square!

**Problem 2:**

The initial number of copies is 500. Because the number triples each year, 3 is the base of the exponent. Multiply the initial number by the exponent. Therefore, $500(3)^t$ = number of copies after $t$ days. Substitute 15 for $t$. $500(3)^{15} = 7174453500$ copies eaten by the monster when he is 15 days old.

**Problem 3:**

Make a factor tree for 86400000. Carefully count the number of each prime number as you write out the prime factorization: $(2 \times 2 \times 2 \times 2 \times 2 \times 2 \times 2 \times 2 \times 2 \times 2) \times (5 \times 5 \times 5 \times 5 \times 5) \times (3 \times 3 \times 3) = 2^{10} \times 5^5 \times 3^3$.

**Problem 4:**

First, calculate the part in parentheses. Be careful of the minus sign in the top of the equation. $(-3.14)^3 = -30.96$. Notice that the answer is negative. Subtract: $3.14 - (-30.96) = 34.1$. Now multiply: $(-5) \times (34.1) = -170.5$. Now evaluate the parentheses in the denominator. $3^2 - 2^3 = 9 - 8 = 1$. Finally, $\frac{-170.5}{1} = -170.5$.

**Problem 5:**

Evaluate each of the exponents carefully before performing any other calculations. $-3^5 = -243$. $(-4)^4 = 256$. $(-1)^1 = -1$. And in the denominator: $2.25^2 = 5.0625$ and $3^4 = 81$. Now evaluate the expression in the parentheses. $(-243) + (256) = 13$. Multiply this by $-2.1$ before subtracting $-1$. $(-2.1) \times (13) - (-1) = -27.3 - (-1) = -26.3$. Working in the denominator, multiply the two numbers with exponents before subtracting. $(5.0625) \times (81) = 410.063$. Now subtract: $2.5 - 410.063 = -407.563$. Now you can divide the expression on the top by the expression on the bottom. And the final answer is $^{-26.3}/_{-407.563} = 0.06453$.

**Problem 6:**

To simplify, combine all of the terms with the same bases. First we reduce all composite numbers to their prime factorization $10^{20} = (2 \times 5)^{20} = 2^{20} \times 5^{20}$. Combine all the powers of 5 $5^3 5^2 5^3 5^{20} = 5^{3+2+3+20} = 5^{28}$. Now combine all the powers of 2 $2^{20} 2^{14} = 2^{34}$. The product of these two terms is our final answer, $5^{28} 2^{34}$.

**Problem 7:**

Combine all the terms with the same bases. $2^5 \times 2^4 = 2^{5+4} = 2^9$ and $3^2 \times 3^5 = 3^{2+5} = 3^7$. Put these two simplified parts together: $2^9 \times 3^7$.

**Problem 8:**

Remember the rule that says if a negative number is raised to an odd exponent, then the answer is still negative. 2,357 is odd, so the answer must be negative.

**Problem 9:**

Count the number of times $5^4$ occurs in the expression. There are five sets of $5^4$, so multiply $5^4$ by 5. The expression is equal to $5(5^4) = 5^5$.

**Problem 10:**

The term $(-6)^9$ occurs 6 times in the expression, so multiply is by 6. The expression is equal to $6(-6)^9$. Because there is no negative sign on the 6, we cannot combine it with the $(-6)^9$.

**Problem 11:**

First, combine the second two terms. $8^5 \times 8^3 = 8^{5+3} = 8^8$. $\frac{1}{8}$ can be written as $8^{-1}$. Again, combine terms by adding the numbers in the exponents $8^{5+3-1}$. So the final answer is $8^7$.

**Problem 12:**

Add all of the exponents for (–7). Adding the exponents you have seven 5's and a 21, so $7 \times 5 + 21 = 56$. So you have $(-7)^{56} \times \frac{1}{(-7)^{13}}$. Because the bases are the same for each term, you can subtract the bottom exponent from the top exponent. $56 - 13 = 43$. So the end result is $7^{43}$.

**Problem 13:**

Add the exponents of all the terms that are multiplied together. $13^2 \times 13^3 \times 13^5 = 13^{10}$. Combine the exponents of the terms that are divided. $13^4 \times 13^2 = 13^6$. Now subtract the bottom exponent from the top. $13^{10} \div 13^6 = 13^{10-6} = 13^4$.

**Problem 14:**

You need to find all of the hidden bases in this problem. Hopefully you can make it so they all have the same bases. $(4)^{-2} = (2^2)^{-2} = 2^{-4}$. $(16)^3 = (2^4)^3 = 2^{12}$. $(8)^{-6} = (2^3)^{-6} = 2^{-18}$. $(-1,024)^2 = (2^{10})^2 = 2^{20}$. Now, you have 2 raised to a bunch of different powers. This problem should look familiar now. Add all of the exponents together to simplify: $2^{(-4)+12+(-18)+20} = 2^{10}$.

**Problem 15:**

Try to find all of the hidden bases for this problem by factoring the bases. $(-8)^4 = ((-2)^3)^4 = (-2)^{12}$. $(9)^2 = (3^2)^2 = 3^4$. $(16)^{-3} = (2^4)^{-3} = 2^{-12}$. $(27)^4 = (3^3)^4 = 3^{12}$. Now, be careful! You can only combine the terms with the same base. You may think that –2 and 2 are different bases; however, –2 is the same as (–1)(2), so you can still combine them. Combining the terms with 2 in the base, you get $(-1)^{12} \times (2)^{12} \times 2^{-12} = 2^0 = 1$. Now combine the terms with 3 in the base. $3^4 \times 3^{12} = 3^{16}$. Now write the simplified terms all together: $3^{16} \times 1 = 3^{16}$. You can check this answer by multiplying $3^{16}$ out on a calculator and the original expression.

**Problem 16:**

$(-1/81) = (-3^{-4})$. Subtract the exponents when dividing: $3^{9-(-4)} = 3^{13}$. You cannot simplify the other terms, so the final answer is simply (no pun intended): $(3^{13}) \times (-5)^{12} \times (0.2)^{-4}$.

**Problem 17:**

Factor 210: $2 \times 5 \times 3 \times 7$. All of these are prime numbers, so the $210 = 2 \times 3 \times 5 \times 7$.

**Problem 18:**

Find all of the hidden bases in each term. Write the bases as a product of primes. $6^5 = (2 \times 3)^5$. $63 = 7 \times 9 = 7 \times 3^2$. Write everything out that you now know: $2^5 \times 3^5 \times 7 \times 3^2 \times 2^2 \times 7^6$. Now combine similar bases to simplify fully: $2^7 \times 3^7 \times 7^7$. You can't combine the bases any further, so you have found the fewest bases possible.

**Problem 19:**

Write the bases as a product of primes:

$$(6)^2 = 2^2 \times 3^2$$

$$70 = 2 \times 5 \times 7$$

$$22 = 2 \times 11$$

$$35^5 = 5^5 \times 7^5$$

Now combine bases: $(-2)^4 \times 2^2 \times 2 \times 2 = 2^8$

$$5 \times 5^5 = 5^6$$

$$7 \times 7^5 = 7^6$$

Writing all together: $2^8 \times 3^2 \times 5^6 \times 7^6 \times 11$. Again, each of the bases is prime, so you cannot simplify this any further.

# Chapter 8

**Problem 1:**

Write the prime factorization of 121. $121 = 11^2$. Luckily, this is all you need to do. 11 is the square root of 121.

**Problem 2:**

Think of this as the square root of 1 divided by the square root of 36. You know that the square root of 1 is 1 and that the square root of 36 is 6, so the answer is 1/6. Again, you can square $\frac{1}{6}$ to check this.

**Problem 3:**

$315 = 3^2 \times 5 \times 7$. Take half of the exponent of $3^2$ and leave the other numbers in the radicand. $3 \times (5 \times 7)^{1/2} = 3(35)^{1/2}$.

**Problem 4:**

First, remember that 27 is between $5^2$ and $6^2$. Because 27 is closer to 25 than to 36, the answer will probably be closer to 5 than 6. Start guessing by choosing 5.2. $5.2^2 = 27.04$. Luckily, this first guess was close enough to 27. So the square root of 27 is approximately equal to 5.2.

**Problem 5:**

Find the prime factorization of 125 first. $125 = 5^3$. Because you are finding the cube root, you need three copies of a number, in this case 5. So the cube root of 125 is 5. Multiply out $5 \times 5 \times 5 = 5^3$ to check.

For the 4th root of 16, again find the prime factorization of 16 and look for 4 copies of the same number. $16 = 2^4$. So the fourth root of 16 is 2.

The prime factorization of 16,807 is $7^5$. Because you have 5 copies of 7, you know that the 5th root of 16807 is 7.

**Problem 6:**

The prime factorization of 250 is $2 \times 5^3$. You have three copies of 5, but only one copy of 2. Leave 2 in the radicand, but multiply it by 5. $250^{1/3} = 5 \times 2^{1/3}$.

The prime factorization of 144 is $2^4 \times 3^2$. $144^{1/4} = 2 \times 9^{1/4} = 2 \times 3^{1/2}$.

Separate the root into the numerator and the denominator and find the prime factorization of each. $69 = 3 \times 23$ and $243 = 3^5$. Because you have five copies of 3 in the denominator, but no sets of five of any number in the numerator, the solution is $\dfrac{69^{\frac{1}{2}}}{3}$.

**Problem 7:**

In order to rationalize the denominator, multiply the top and bottom by the square root of 3. $4 \times 3^{1/2} = (3^{1/2} \times 3^{1/2}) = 4 \times \dfrac{3^{\frac{1}{2}}}{3}$.

**Problem 8:**

Multiply by $12^{1/2}$. The solution is $-\dfrac{12 \times 12^{\frac{1}{2}}}{12} = 12^{\frac{1}{2}}$.

**Problem 9:**

Remember the order that you multiply in: $(5^{1/2})(5^{1/2}) - (5^{1/2})(3^{1/2}) + (3^{1/2})(5^{1/2}) - 3 = 5 - 3 = 2$.

**Problem 10:**

Multiply the top and bottom by the conjugate of the denominator.

$$\frac{5\left(17^{\frac{1}{2}}-19^{\frac{1}{2}}\right)}{\left(17^{\frac{1}{2}}+19^{\frac{1}{2}}\right)\left(17^{\frac{1}{2}}-19^{\frac{1}{2}}\right)}=\frac{5\left(17^{\frac{1}{2}}-19^{\frac{1}{2}}\right)}{(17-19)}=\frac{5\left(17^{\frac{1}{2}}-19^{\frac{1}{2}}\right)}{-2}.$$

**Problem 11:**

Raising something to the $\frac{1}{2}$ power is the same as taking the square root, so the solution is simply 5.

**Problem 12:**

Raising a number to the $\frac{1}{3}$ is the same as taking the cube root, so the solution is 6.

**Problem 13:**

First, evaluate $25^{1/2}$, then cube the solution. We already now that the square root of 25 is 5, so $5^3 = 125$ is the solution.

**Problem 14:**

First, evaluate $64^{1/3}$. Then, take that solution to the fourth power. $64^{1/3}$ is 4. $4^4 = 256$.

**Problem 15:**

Ignore the minus sign in the exponent until the end. $25^{1/2} = 5$. $5^3 = 125$. Now, since the exponent is negative, the solution is simply $\frac{1}{125}$.

**Problem 16:**

From Problem 14, we know that $64^{4/3} = 256$. Because the exponent is negative in this problem, the answer is $\frac{1}{256}$.

**Problem 17:**

To rewrite this expression, first evaluate the exponent $64^{5/6}$. $64^{5/6} = 2^5 = 32$. Now, multiply this by the exponent $-5/2$. $32 \times (-5/2) = -80$. So the expression is equivalent to $8^{-80}$.

**Problem 18:**

To change to exponential notation, rewrite all of the exponents in the denominator in the numerator, but with inverse exponents. The solution is: $(4^{-2})(3^7)(5)(7^{-4})(11^{-3})$. Even though the two look pretty different, they are equivalent!

**Problem 19:**

To change to fraction notation, move all of the numbers with negative exponents into the denominator and give them positive exponents. The expression is equivalent to $\dfrac{\left(2^4\right)\left(21^{13}\right)}{\left(3^5\right)\left(9^7\right)}$.

# Chapter 9

**Problem 1:**

Take the number farthest to the left and put it to the left of the decimal point. Then, count the number of decimal places. Marvin has $2.345 \times 10^{21}$ hamsters!

**Problem 2:**

Move the decimal point 23 spots to the right since the exponent is positive. Count carefully! Avogadro's number is 620,000,000,000,000,000,000,000.

**Problem 3:**

Because the exponent is negative, move the decimal point to the left 11 places. G = 0.0000000000667 (a very small number!).

**Problem 4:**

Count the number of decimal places until you get to the right of the 1, the first non-zero number. The mass of a neutron is $1.67 \times 10^{-24}$ grams.

**Problem 5:**

Calculate the amount of money after 10 years for each account.

1) The principal is $10,000, the interest rate is .1, and the time is 10 years. The amount of money after 10 years is $10,000(1 + .1)^{10} = \$25,937.40$.

2) The principal is $8,000, the interest rate is .15, and the amount of time is 10 years. The amount of money after 10 years is $8,000(1.15)^{10} = \$32,364.50$. The second offer is definitely much better (assuming you want more money!).

**Problem 6:**

$1,500,000 is the principal and the interest rate is 3.5%, so the endowment will generate $0.035 \times \$1,500,000 = \$52,000$ per year. Dividing this up into monthly chunks, $52,000 per year = $\dfrac{\$52,000}{\text{year}} \cdot \dfrac{1 \text{ year}}{12 \text{ months}} = \dfrac{\$4,333.33}{\text{month}}$. You can spend up to $4,333.33 per month.

**Problem 7:**

In this situation, $P = 435{,}000$, $r = .105$, $Y = 30$. Plug these numbers into the variables in the equation and calculate carefully. Remember to follow the order of operations and to start with the innermost parentheses.

$$PMT = \frac{P \times \left(\dfrac{r}{12}\right)}{\left[1 - \left(1 + \dfrac{r}{12}\right)^{-12Y}\right]} = \frac{435{,}000 \times \left(\dfrac{0.105}{12}\right)}{\left[1 - \left(1 + \dfrac{0.105}{12}\right)^{-12 \times 30}\right]} = \frac{435{,}000 \times (0.00875)}{\left[1 - (1 + 0.00875)^{-360}\right]} =$$

$$\frac{3{,}806.25}{[1 - 0.04344]} = \frac{3{,}806.25}{[0.956561]} = 3{,}979.12$$

**Problem 8:**

The area is the length times the width and since it's a square, the area is simply $2.5^2 = 6.25$.

**Problem 9:**

Because the area of a square is the side-length squared, take the square root of the area to find the side-length. Did you follow that? If $x$ is the side length, then the area is $x^2$ which is 49. Therefore, $x = 49^{1/2} = 7$.

**Problem 10:**

You can use a calculator to find that the square root of 802 is a little bit more than 28. So you can check all of the prime numbers up until 28: 2, 3, 5, 7, 9, 11, 13, 17, 19, 23 to see if they divide into 802.

**Problem 11:**

The square root of 782 is a little more than 27, so you can check all of the prime numbers up to 27 to see if they divide into 782. We end up with 782 having the factors 2, 17, and 23.

# Chapter 10

**Problem 1:**

Check each term to see if it is a valid monomial. $x^2$ is a valid monomial because the power is a positive integer. The next term is also valid because $x$ is raised to the first power. Even though the constant is not raised to an integer power, it doesn't matter;

this is still valid. For the next term, remember that the square root of $x$ is the same as $x$ raised to the one-half power, which is not an integer power. The next term is valid, but since $x^{1/2}$ is invalid, the entire expression is not a polynomial.

**Problem 2:**

All of the terms are raised to positive integer powers. The last term, 1, is the coefficient for $x$ raised to the zeroth power. All of the terms are valid monomials, so this expression is a polynomial.

**Problem 3:**

If you think of this as $(3.94 + 5.32)x$, then you can see how easy it is to add the coefficients. The sum is equal to $9.26x$.

**Problem 4:**

Subtract the coefficients, or add a minus 563.3. The sum equals $-329.1y$.

**Problem 5:**

Multiply the coefficients. $4 \times 7 = 28$. Altogether, the expression is equal to $28x^8$.

**Problem 6:**

Multiply the coefficients and follow the rule of exponents to find the final product. $5 \times (-9) = -45$. $(x^3)(x^{20}) = x^{3+20} = x^{23}$. $(5x^3)(-9x^{20}) = -45(x^{23})$.

**Problem 7:**

Use the Distributive Law and the rules of multiplying exponents. $x^4(7x^5) - x^4x^2 = 7x^9 - x^6$.

**Problem 8:**

Watch the signs in this problem. $(-5x)(3x^3) + (-5x)(-9x^5) = (-15)x^4 + 45x^6$.

**Problem 9:**

Follow the rules of FOIL carefully, starting by multiplying out the first terms in each expression. $(7)5x^2 + (7)8x^3 + (3x)(5x^2) + (3x)(8x^3) = 35x^2 + 56x^3 + 15x^3 + 24x^4$.

**Problem 10:**

Use FOIL carefully and watch out for the minus sign. $(8x)(11x^4) + (8x)(8x^3) + (-5x^7)(11x^4) + (-5x^7)(8x^3) = 88x^5 + 64x^4 - 55x^{11} - 5x^{10}$.

**Problem 11:**

Replace $x$ with 5 in the equation: $2(5)^3 = 2(125) = 250$.

**Problem 12:**

You can simplify the expression first, or simply plug in 4 for each value of $x$. Either way, you will get the same answer (as long as you do it correctly!). $(-5)(4)(3(4^3) - 9(4^5))$ $= (-20)(3(64) - 9(1,024)) = (-20)(192 - 9,216) = (-20)(-9,024) = 180,480$.

**Problem 13:**

In English, this is eleven more than eleven times a number $x$.

**Problem 14:**

Simplify the expression first to make it easier: $(-15)x^4$. Minus fifteen times a number $x$ taken to the fourth power.

**Problem 15:**

$F = \frac{9}{5}0 + 32 = 32$. So 0°C in Fahrenheit is 32°.

# Chapter 11

**Problem 1:**

Add 7 to both sides of the equation because adding 7 is the inverse of subtracting 7: $x - 7 + 7 = 4 + 7 = 11$. Therefore, $x = 11$.

**Problem 2:**

First, simplify the left side of the equation: $17 - 6 = 11$. Now, apply the inverse operation of adding 2: $x + 2 - 2 = 11 - 2 = 9$. So $x = 9$.

**Problem 3:**

You need to find the inverse operation of $\frac{5}{29}$, which would be multiplying by its reciprocal: $\frac{29}{5}$. $(\frac{29}{5})(\frac{5}{29})z = (\frac{29}{5})5 = 29$. Therefore, $z = 29$. Multiplying by the reciprocal is the same as dividing by the number, so either operation is the inverse.

**Problem 4:**

$x$ is multiplied by 6, so you need to divide by 6 or multiply by $\frac{1}{6}$. I'll do the latter since multiplication is easier for me. $\frac{1}{6} \times \frac{3}{7} = \frac{3}{42}$ and $(\frac{1}{6})6x = x$ so $\frac{3}{42} = x$.

**Problem 5:**

To peel this onion of a problem, apply the inverse operation of adding $\frac{5}{29}$, which is subtracting $\frac{5}{29}$. $\frac{5}{29} - \frac{5}{29} + 4a = 5 - \frac{5}{29} = \frac{140}{29}$. Now multiply by $\frac{1}{4}$. $(\frac{1}{4})4a = a$ and $\frac{140}{29} \times \frac{1}{4} = \frac{35}{29}$ so $a = \frac{35}{29}$. The key here is remembering to subtract first, *then* multiply by the reciprocal.

**Problem 6:**

Subtract 1 from both sides, then multiply by –2 because –2 is the inverse of $-\frac{1}{2}$. $(5 – 1)(–2) = –8 = x$.

$2x + 7 + 4x = 18$

Solving for $x$, first combine terms with $x$. Then subtract 7 from both sides: $6x + 7 – 7 = 18 – 7 = 11$. Now divide by 6. $x = \frac{11}{6}$.

$5 – 3x = 2$

The easiest way to think of this problem is probably to add $3x$ to both sides, then subtract 2. $5 – 2 – 3x + 3x = 2 – 2 + 3x$. $3 = 3x$. Dividing by 3, we get $x = 1$.

$3x – 9 = 9$

Add 9 to both sides, then divide by 3. $\frac{1}{3} \times (3x – 9 + 9) = x$ and $\frac{1}{3}(9 + 9) = 6$ so $x = 6$

**Problem 7:**

Combine the terms with a on the left by adding the coefficients. $a + 4a = 5a = 17 – a$. Add a to each side. $5a + a = 17 – a + a$. Now divide by 6. $(\frac{1}{6})6a = \frac{1}{6}17 = \frac{17}{6} = a$.

**Problem 8:**

You can collect the z terms on either side of the equals sign. It might be easiest to add $\frac{z}{6}$ to both sides, but the choice is up to you. $3 – \frac{z}{6} + \frac{z}{6} = 8 + z + \frac{z}{6}$. Now subtract 8 from both sides. $3 – 8 = 8 – 8 + \frac{7}{6}z$. Now multiply by the reciprocal of $\frac{7}{6}$. $\frac{6}{7}(–5) = –\frac{30}{7}$ and $(\frac{6}{7})(\frac{7}{6})z = z$, so $z = –\frac{30}{7}$.

**Problem 9:**

This may look confusing because there are so many variables and you won't get a plain number in the end. Your end result will be $x$ on one side with the other letters on the other side of the equals sign. Start by subtracting z from both sides. Then, divide by 10. $\frac{1}{10}(5a – \frac{b}{4} – z) = (\frac{1}{10})10x + z – z = x$. So, $\frac{1}{2}a – \frac{b}{40} – \frac{z}{10} = x$.

**Problem 10:**

This one will be a little messier because $a$ is on both sides of the equation. Try to get all of the $a$'s on the right side, and all of the other variables on the left side. $–3z – yx^3 = 8a + \frac{5a}{15}$. Now combine the terms with $a$ in them and apply the inverse operation to both sides. $\frac{15}{125}(–3z – yx^3) = (\frac{15}{125})(\frac{125}{15})a$. So $a = \frac{15}{125}(–3z – yx^3)$, which is a messy but true answer.

## Problem 11:

For this problem, all of the terms are $x^2$, so you just need to add all of the coefficients. Watch out for the plus and minus signs! $1 + (-2) + 3 + (-4) + 5 + (-6) + 7 + (-8) = -4$. So the expression is equivalent to $-4x^2$.

## Problem 12:

First off, organize all of the terms with $x$ raised to the same power together and then add the coefficients of like powers. $32x^7 + 4x^6 + \frac{3}{4}x^5 + 15x^4 + 5x^2 + (\frac{1}{2}x^1 - 12x^1 - 3x^1)$ $= 32x^7 + 4x^6 + \frac{3}{4}x^5 + 15x^4 + 5x^2 + (-\frac{29}{2})x^1$.

## Problem 13:

When simplifying and combining like terms, make sure that both variables in the term are raised to the same powers. Remember that a variable or a number raised to the zero power is just one, so you can simplify some of the terms. $17x + 5x - 2xz^1 + \frac{5}{2}xz^1 + xz^3 - 3xy^1 + 10xz^4 + zx^6 = 22x + \frac{1}{2}xz^1 + xz^3 - 3xy^1 + 10xz^4 + zx^6$. You cannot simplify this any further because all of the terms are different.

## Problem 14:

First, simplify all of the terms where a number or variable is raised to the zeroth power. $5^0 = 1$ and $xz^0 = x$ and $17xy^0 = 17x$. There are two terms with $z^6$ that you can combine. $x^1$ is the same as $x$, so you can combine it with the other $x$ terms. Finally, you should get $100,001 - z^6 + 15x + 11xz + xy^4$.

## Problem 15:

In the equation, put a $(-2)$ wherever you see an $x$. Then, simplify the expression and see if it works. $2 + 12(-2) = (-2)^5 + (-2)^4 + (-2)^3 + (-2)^2 + (-2)^1$. Simplify: $2 - 24 = -32 + 16 - 8 + 4 - 2$. Now just add and subtract; I know you can do it! $-22 = -22$. Because this is a true statement, $x = -2$ is a solution.

## Problem 16:

Plug in the respective numbers for each variable. Make sure you put the right number in the right place! $\dfrac{40 + 2(20)^2}{10} = -18(5) + (\frac{8}{20})((5)(20))^{1/2} + 364$. Now multiply: $(84) = -90 + 4 + 364 = 278$. Unless pigs are flying and mathematics has changed drastically, 84 does not equal 278. This means that $x = 5$, $y = 20$ is *not* a solution to the equation. Try again!

# Chapter 12

**Problem 1:**

You can write out the number of each variable and then cancel to solve the problem. However, since the variables $b$ and $c$ are only in the top and bottom respectively, you don't need to write them out because they won't cancel with anything, making your job a little easier. $\dfrac{70a \times a \times a \times b^2}{150a \times a \times c^7}$. Two $a$'s cancel out on the top and the bottom and $^{70}/_{150}$ reduces to $^7/_{15}$. The final answer is $\dfrac{7a \times b^2}{15c^7}$.

**Problem 2:**

First, put the negative exponent into the top. Then, write out the number of variables and see what cancels. Even though the numbers are separated into two fractions, combine it all into one to simplify. $\dfrac{45x \times x \times x \times x \times z \times z \times z \times z \times z \times y \times y}{66 \times x} \cdot \dfrac{11y}{270x \times y \times y \times y \times z \times z \times z \times z \times z}$. Simplify to $^1/_{36}$. All the variables cancelled out!

**Problem 3:**

Because all the exponents are positive, move both variables to the opposite position and make the exponent negative. $\dfrac{24y^{-4}}{43x^{-2}}$.

**Problem 4:**

Cross-multiply using parentheses and carefully distributing. $3(7a) = 6a(^1/_2 + 2)$. Distributing: $21a = 3a + 12a = 15a$. Move all of the $a$'s onto one side of the equation to get $6a = 0$, so $a = 0$.

**Problem 5:**

Using the equation given, $a = 900$ and $b = 300$ (or vice versa, really, but it doesn't matter!). Plug these numbers into the equation and solve for $T$. $^1/_{900} + ^1/_{300} = ^1/_T = ^4/_{900}$. Cross-multiplying, we see that $^{900}/_4 = 255 = T$. So it will take 225 minutes to fill the pool if Trillian and Arthur work together. For Trillian, this is only a little bit faster, but for Arthur this is amazingly fast!

**Problem 6:**

We can think of this problem as Dagny and the train moving along a line segment of length 120, Dagny starting at 120, the train starting at 0, and they're heading toward each other. Oceanside is located at 50.

Because Dagny starts at 120 and is headed toward the train at a speed of 80, her position will be D = 120 - 80 $t$. The train starting at 0 and moving toward Dagny will have a position of 0 + 50 $t$. We want to find where the train is when Dagny reaches the it. If it's position is less than 50, Dagny has made it in time.

Symbolically, Dangy reaches the train when D = T, so we're interested in where the train is at time $t$ such that D = T. Substituting in the expressions with $t$'s in them for D and T, we know that when Dagny arrives at the train, 120 – 80 $t$ = 50 $t$. Solving for $t$, we first add 80 $t$ to each side, ending up with 120 = 130 $t$. Then we divide both sides by 130, reaching the solution $120/130$ = $t$, or $t$ is about 0.923 hours.

Finally, we want to plug in this value of $t$ to the equation for the train's position.

$$T = 50 \cdot \left(\frac{120}{130}\right) = \frac{6000}{130} = \frac{600}{13}. \text{ Because } \frac{600}{13} \cdot 13 = 600 < 650 = 50 \cdot 13, \text{ T} = \frac{600}{13} < 50 \text{ and}$$

Dagny reaches the train in time.

# Chapter 13

**Problem 1:**

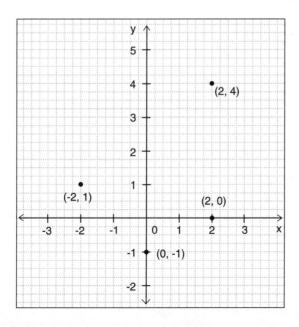

**Problem 2:**

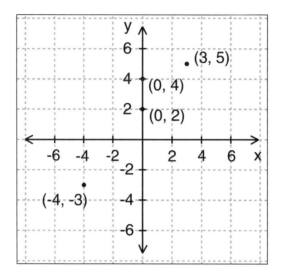

**Problem 3:**

Plug in a variety of *x* values, such as –2, –1, 0, 1, and 2. This will give you a very clear idea of what the line looks like.

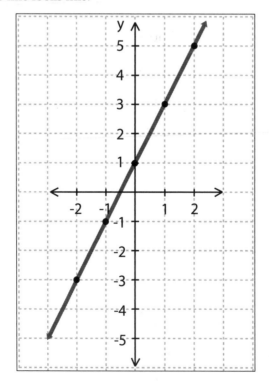

**Problem 4:**

Once again, try to graph a variety of $x$ values so you can clearly see the shape of the graph. You will want to make a table to keep everything organized:

| $x$ | $y$ |
|---|---|
| $-4$ | 12 |
| $-3$ | 6 |
| $-2$ | 2 |
| $-1$ | 0 |
| 0 | 0 |
| 1 | 2 |
| 2 | 6 |
| 3 | 12 |
| 4 | 20 |
| 5 | 30 |

Now plot these points. You should get a u-shape with a very shallow bottom. The bottom of the U-shape is actually negative between −1 and 0, which you can see if you plot extra points in that interval.

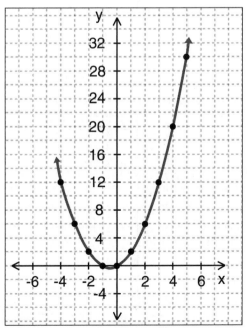

**Problem 5**:

You can find the distance using the distance formula. $D = ((7 - 3)^2 + (-1 - 2)^2)^{1/2} = (16 + 9)^{1/2} = 5$. The midpoint formula gives the following result: $(3 + 7) \div 2 = 5$ and $(2 + -1) \div 2 = \frac{1}{2}$. So the midpoint is at $(5, \frac{1}{2})$.

# Chapter 14

### Problem 1:

The angle should measure 120°. If you don't actually have a clock you can still figure out the answer. A clock is a circle divided into twelve sections. The angle between two numbers with the center of the clock as the vertex is $360 \div 12 = 30$. There are four sections between 12 and 4, so the angle is $30 \times 4 = 120$ degrees.

# Chapter 15

### Problem 1:

By definition, a triangle with at least two sides of the same length is an isosceles triangle. If all of the sides were of length 4, then it would still be an isosceles triangle!

### Problem 2:

This triangle has all three sides of length $a$. By definition, a triangle with at least two sides of the same length is an isosceles triangle. Also by definition, a triangle with three sides equal is equilateral. Thus this triangle is both isosceles and equilateral.

### Problem 3:

Recall the both definitions. A kite has no parallel sides, but two pairs of sides with equal lengths. A rhombus has two pairs of sides with equal lengths also, but its opposite sides are parallel. Therefore, no kite is a rhombus.

### Problem 4:

Using the British definition of a trapezoid, which is the same as the American definition of a trapezium (confusing, I know), we know that a trapezoid does not have any parallel sides. (It wouldn't make much sense to use the American definition of a trapezoid, since no trapezoids would be a kite). A kite is a British trapezoid with two pairs of sides of equal lengths. By definition, every kite is a British trapezoid, but some trapezoids are not kites. That means that some trapezoids do not have two pairs of sides of equal lengths.

**Problem 5:**

Simply use the formula (it works even for really-big-number-a-gons). $(180 - 2) \times 180 = 178 \times 180 = 32{,}040$ degrees in each angle.

**Problem 6:**

You can use the formula again, but you'll have to work backwards. If $n$ is the number of sides, then $(n - 2) \times 180 = 720$. Divide by 180 and add 2 to both sides (in that order). $720 \div 180 = 4$. $4 + 2 = 6 = n$. The polygon has six sides, so it is a regular hexagon.

**Problem 7:**

You can draw a picture of the rectangular field and label the short sides 40 ft. and the longer sides 320 ft. You have two sides of length 40 and two sides of length 320. Adding these together you get $40 + 40 + 320 + 320 = 720$. The farmer will need 720 feet of fence to enclose the fields.

**Problem 8:**

A rhombus has four equal side lengths, so its perimeter will be $4 \times 4 = 16$. The rhombus will have a perimeter of 16 centimeters.

**Problem 9:**

In this case, $c = 13$ and one of the legs, say $a$, is 5. We need to use Pythagorean Theorem to find $b$. $13^2 = 5^2 + b^2$, so $b^2 = 13^2 - 5^2 = 144$. The square root of 144 is 12, so the other leg of the triangle has a length of 12 centimeters.

# Chapter 16

**Problem 1:**

Area = base × height. In this case, Area = $0.5 \times 249 = 124.5$.

**Problem 2:**

The area of a parallelogram is base times height, so we need to identify the base and height in this diagram. A base is a side of a parallelogram and the height a measurement perpendicular to it that goes from the line the base lies in to a point on the line the opposite side lies in. With this one it's pretty easy. The base is 25 units long, and the height is 10 units tall. (Ignore the slanty side. That number is there to confuse you.) Therefore, the area of the parallelogram must be $25 \times 10 = 250$ square units.

**Problem 3:**

The area of a trapezoid is the average of the two bases times the height, or $\frac{(b_1 + b_2)}{2} \cdot h$.

Because which base is called $b_1$ and which base is called $b_2$ is immaterial, we choose $b_1 = 5$ and $b_2 = 9$. $h = 5$, so the total area is $\frac{(5+9)}{2} \cdot 5 = 35$ square units.

**Problem 4:**

First find the length of the third side. $13^2 = 12^2 + b^2$. $b = (13^2 - 12^2)^{1/2} = 5$. So the lengths of the two sides are 12 and 5, as are the lengths of the base and height of the triangle. So the area is $\frac{1}{2} \times 12 \times 5 = 30$. The area is 30 square meters and the unknown leg is 5 meters long.

**Problem 5:**

The circumference of a circle is $C = 2\pi r$, so for this circle, $\pi = 2\pi r$. Therefore, $r = \frac{1}{2}$.

**Problem 6:**

You will travel the circumference of the wheel each time it rotates once. The diameter is 26 inches, so the circumference and the distance you travel is $26\pi$ inches.

**Problem 7:**

First, solve for the radius from the circumference. $2\pi r = 14$ so $r = \frac{7}{\pi}$. The area is $\pi r^2$. $(\frac{7}{\pi})^2 \times \pi = \frac{49}{\pi}$. The area is $\frac{49}{\pi}$ square inches.

**Problem 8:**

Once again we're going to use the idea of removing area from a larger figure, this time combined with breaking down some areas into smaller triangles. We're going to use the area formulas for a circle $\pi r^2$ and for a rectangle $l \cdot w$.

The area of the outer circle is $\pi 8^2 = 64\pi$. The area of the small circle on the left is $\pi 1.5^2 = 2.25\pi$, and the area of the small circle on the right is $\pi 1^2 = \pi$. We're going to break the "mouth" region of the figure into three rectangles with two of length 4 and width 2, and one of length 2 and width 6. The total area of these three rectangles is $2 \cdot 4 + 2 \cdot 4 + 6 \cdot 2 = 28$.

Finally, we calculate the area of the shaded region as square units, or about 166.8 square units.

**Problem 9:**

The volume of the cube-shaped chair is the side length cubed. $2^3 = 8$. So you will need 8 cubic feet of stuffing to reanimate the chair.

**Problem 10:**

Because the volume is the product of the side lengths, you can find the third side length by dividing the volume by the two known side lengths. $\frac{30}{2\times3}$ = 5. The surface area is then 2(2 × 3) + 2(2 × 5) + 2(3 × 5) = 62.

**Problem 11:**

The area of the base is 24 × 24 = 576. The height is 8 feet, so the volume of concrete needed is $\frac{1}{3}$ × 576 × 8 = 1536. 1536 cubic feet of concrete are needed.

**Problem 12:**

The volume of the tank is found from the formula $V = \pi r^2 h$ where $r$ = 2 and $h$ = 5. So $V$ = 20$\pi$. Multiply the weight of one cubic foot of water by the number of cubic feet. 62.5 × 20$\pi$ = 1,250$\pi$. The tank weighs 1,250$\pi$ pounds, which is well over 3,750 pounds!

**Problem 13:**

Setting the volumes equal we get: $\frac{1}{3}\pi r_1^2 h = \frac{1}{3}\pi r_2^2 h$. Multiply both sides by 3 and divide by $h$ to get $\pi r_1^2 = \pi r_2^2$. This shows that the area of the bases are equal, though the radius is different for each figure. So the ratio of the surface areas is 1.

# Chapter 17

**Problem 1:**

If you drew them correctly, both squares will have all equal side lengths, and, since they are squares, they will each have four right angles. Because all of the angles and side lengths are equal, they must be congruent!

**Problem 2:**

Are all rhombuses similar to each other?

In order to be similar, two rhombuses would have to have congruent angles and the side lengths would be scaled up or down by some scale factor. However, two rhombuses do not have to have congruent angles, so they are not all similar.

**Problem 3:**

To find the scale factor, divide the length of a side of the larger pentagon by the length of a side of the smaller pentagon.

The larger pentagon has side length 12 in and the smaller pentagon has side length 5 in. Thus the scale factor is $^{12}/_5$.

**Problem 4:**

The area of a parallelogram is base times height, so we need to identify the base and height in this diagram. A base is a side of a parallelogram and the height a measurement perpendicular to it that goes from the line the base lies in to a point on the line the opposite side lies in. With this one it's pretty easy. The base is 25 units long, and the height is 10 units tall. (Ignore the slanty side. That number is there to confuse you.) Therefore, the area of the parallelogram must be $25 \times 10 = 250$ square units.

# Chapter 18

**Problem 1:**

The range is the highest number minus the lowest number. In this case, $17 - 2 = 15$ is the range of ages. This means that Mama Snook stopped having children 15 years after she started.

**Problem 2:**

| Value | Frequency | Percentage |
|-------|-----------|------------|
| Brown | 3 | 25% |
| Blue | 3 | 25% |
| Green | 4 | 33.3% |
| Grey | 1 | 8.3% |
| Hazel | 1 | 8.3% |

To find the frequency for each eye color, simply count the number of times the color came up in the given list. The percentage is the frequency divided by 12, since there are twelve pairs of eyes being examined.

**Problem 3:**

Count the number of scores between 80 and 89 first. 4 people got exactly a B. Some students scored higher and got better than a B. 2 scores were higher than 89, so a total of 6 students got a B or better.

**Problem 4:**

You will need to find all of the cars that were driving between 25 and 35 miles per hour. The 5 cars driving speeds of 29, 31, 25, 25, and 33 were driving in this range.

**Problem 5:**

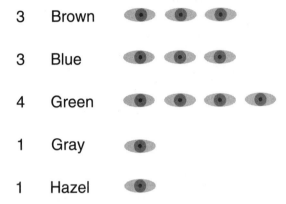

| 3 | Brown |
| 3 | Blue |
| 4 | Green |
| 1 | Gray |
| 1 | Hazel |

**Problem 6:**

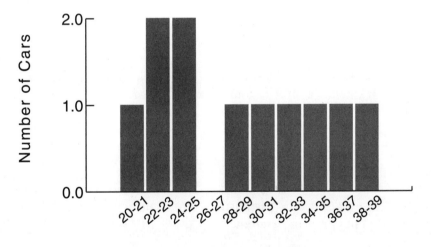

**Problem 7:**

The only difference between the pictograph and the bar graph is, of course, that the bar graph has solid bars instead of pictures showing the number of people with a given eye color.

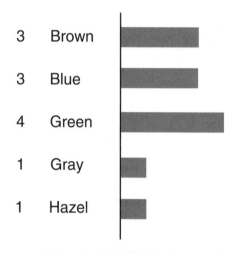

| | |
|---|---|
| 3 | Brown |
| 3 | Blue |
| 4 | Green |
| 1 | Gray |
| 1 | Hazel |

**Problem 8:**

Answers will vary depending on who you know, but it is important to remember two things when making the plot. First, if some of the people you know are younger than 10, you should use a 0 on the left side of the plot to represent the first digit of their age. Second, if you know three or four people of the same age, then you need to make sure to write out the second digit of their ages that many times. For instance, if you know three people who are 14 years old, you should write three 4's on the right side of the plot next to the 1's digit.

**Problem 9:**

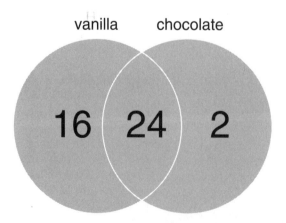

**Problem 10:**

The mean of all runners, rollers, and sleepers is:

$$\frac{5+14+16+16+16+20+21+303}{8} = \frac{416}{8} = 52.$$ The mean for all eight is 52 seconds.

The mean for the six friends is: $\frac{14+16+16+16+20+21}{6} = \frac{103}{6} = 17\frac{1}{6}$. On average,

Carl and Janet's friends took about 17.167 seconds to run 100 meters. We can see that Carl's time raises the entire mean quite a bit! If we looked just at the mean for all eight, we would think Carl and Janet's friends should exercise more!

**Problem 11:**

First, order the data: 2 8 9 10 13 16. The median will be the average of the two middle numbers, since there is an even number of data. The average of 9 and 10 is 9.5 so the median is 9.5. Notice that this median is not an actual datum given in the problem.

**Problem 12:**

Count the number of times each number of pets comes up. Five students have 0 pets, four students have 4 pets, three students have 2 pets, one student has 4 pets, two students have 5, and one has 6 pets. The most common number of pets is 0.

# Chapter 19

**Problem 1:**

Your sample space contains HH and HT. HT is the result we are interested in, so the probability is then $\frac{1}{2}$.

**Problem 2:**

Your sample space contains HHHH, HHHT, HHTT, HHTH, HTHH, HTHT, HTTH, HTTT, THHH, THHT, THTT, THTH, TTHH, TTHT, TTTH, TTTT. Not surprisingly, your denominator will be 16. the desired result occurs four times, so the probability is $\frac{4}{16}$ or $\frac{1}{4}$.

**Problem 3:**

First, let's write out the values in our sample space:

| | | | | | |
|---|---|---|---|---|---|
| 1,1 | 1,2 | 1,3 | 1,4 | 1,5 | 1,6 |
| 2,1 | 2,2 | 2,3 | 2,4 | 2,5 | 2,6 |
| 3,1 | 3,2 | 3,3 | 3,4 | 3,5 | 3,6 |
| 4,1 | 4,2 | 4,3 | 4,4 | 4,5 | 4,6 |
| 5,1 | 5,2 | 5,3 | 5,4 | 5,5 | 5,6 |
| 6,1 | 6,2 | 6,3 | 6,4 | 6,5 | 6,6 |

If we say that the yellow die's results are on the left, the above list tells us that in five of them the purple die result is five without the yellow die's result also being five, so the probability is $5/36$.

**Problem 4:**

There are three face cards (jack, queen, king) in each suit and four suits in the deck, so the probability is $12/52 = 3/13$.

**Problem 5:**

Be careful here; your sample space got smaller when the first card was taken, reducing the number of cards in the deck from 52 to 51. Also, the number of sixes in the deck changed from 4 (one in each suit) to 3, so the probability is $3/51 = 1/17$.

**Problem 6:**

We know that the probability of rolling a 6 is $1/6$, so the probability of *not* rolling a six is $6/6 - 1/6 = 5/6$.

# Index

## A

absolute value, 13-14, 76
abstraction, 108
acute angles, 220-221
addition
    equations, 161-162
    expressions, combining like terms, 147-149
    fractions, 58
        like denominators, 58-59
        unlike denominators, 59-61
    integers, 14-16
additive relationships, 79
algebraic techniques, 175
    cross-multiplication, 178-180
    distance-rate-time problems, 182-188
    rate-of-work problems, 180-182
    simplifying fractions, 176-177
angles, 211
    90°, 219
    acute, 220-221
    complementary, 223
    concave, 226
    convex, 225
    estimating, 219
    measuring, 212-215
        protractors, 215, 219
    obtuse, 221
    paired, 222
        complementary angles, 223
        supplementary angles, 224
    right, 222
    straight, 227

applications
    data analysis, 284
    integral exponents, 94
        exponential notation, 96
        repeated multiplication, 94-96
    proportions, 82
        recipes, 82-84
        science applications, 89
    scale factors, 270
area
    calculating, 248
        circles, 254-257
        parallelograms, 250-252
        rectangles, 248-249
        trapezoids, 252-253
        triangles, 254
    scale factors, 276
    squares, 135
arrays, 135
Associative Law, 36-38

## B

bar graphs, 293
base-ten number system, 126-127
binomials, multiplication with the FOIL method, 150-153
bins, sorting data, 288-289
*Book of Calculation*, 126

## C

calculations
  area, 248
    circles, 254-257
    parallelograms, 250-252
    rectangles, 248-249
    trapezoids, 252-253
    triangles, 254
  percents, 68
    finding percentages, 68-70
    percent change, 70-71
  scale factors
    area, 276
    length, 271-276
  square roots, 109
    imperfect squares, 111-114
    perfect squares, 110-111
    radicals, 114
cards, probability, 310-311
Cartesian plane, 192-193
Cartesian plane coordinates, 192-197
Celsius, converting to Fahrenheit, 155
checking work, solving equations, 172
  multiple variables, 173-174
  single variables, 172-173
circles
  calculating area, 254
    distances in circles, 256
    formula, 257
    pi, 255-256
  circumference, 256
  diameter, 256
coefficients, 146
coin flipping, probability, 304-307
combining like terms
  adding expressions, 147-149
  equations, 164-165
commissions, percents, 74-75
common denominators, 59
common factors, 9
  simplifying fractions, 176

common multiples, 11
Commutative Law, 34-36
complementary angles, 223
complex fractions, 57
concave angles, 226
cones, volume and surface area, 263-264
congruent shapes, 266
conjugates, 117
constant terms, 146
conventions, naming variables, 143-144
conversion factors, 81
conversions
  currency, 85-86
  percents
    to decimals, 64-66
    to fractions, 66-68
  temperature, 155
  units, 87
convex angles, 225
coordinate axes, 193
coordinate plane, 193
coordinates, 194
criteria, rational numbers, 29
cross-multiplication, 51, 178-180
cube roots, 115
  rational exponents, 119
cubes, volume and surface area, 258
currency conversions, 85-86
cylinders, volume and surface area, 262-263

## D

data analysis, 303
  applications, 284
  defining data, 283
  displaying data, 290
    bar graphs, 293
    histograms, 292
    line plots, 297
    pictographs, 290
    stem-and-leaf plots, 294-295
    Venn diagrams, 295-297
  frequency tables, 285-287

measures of central tendency, 298
   mean, 298-300
   median, 300-301
   mode, 301
  probability
   chances of event not happening, 311
   independent events, 304-311
  qualitative versus quantitative, 284-285
  sorting data, 287
   bins, 288-289
   outliers, 289-290
decimals, 20
  decimal places, 20
  fractions and, 53-54
   repeating decimals, 55-56
   terminating decimals, 54-55
  number lines, 26-27
  pattern, 20
  percents and, 64-66
  rational numbers, 22
  real numbers, 21
degree measurement of an angle, 214
degrees, polynomials, 146
denominator (fractions), 24
  like denominators, 58-59
  lowest common denominators, 59
  rationalizing square roots, 116
   single square roots, 116-117
   two square roots, 117-118
  unlike denominators, 59-61
density, proportions, 89
diameter (circles), 256
dice, probability, 307-309
dilation, 268
"dirt equations," 182
displaying data, 290
  bar graphs, 293
  histograms, 292
  line plots, 297
  pictographs, 290
  stem-and-leaf plots, 294-295
  Venn diagrams, 295-297

distance measurements, 201
distance-rate-time problems, 182-188
Distributive Law, 38
  minus sign, 39-41
  multiplying expressions, 150
  standard problems, 38
Distributor, The (Distributive Law), 38
divisible (factoring numbers), 6
division
  equations, 162-163
  fractions, 56-58
  integers, 16
  integral exponents, 103
  order of operations, 33-34

# E

endowments, financial math, 132
equations, 157
  addition/subtraction, 161-162
  defined, 158-160
  graphing, 197-200
  inverse operations, 160
  more than one variable, 165-167
  multi-step equations, 163
   combining like terms, 164-165
   more than one operation, 163
  multiplication/division, 162-163
  one operation, 160-161
  solutions, 159
   checking work, 172-174
   more than one solution, 167-168
   multiple variables with various powers,
    169-171
   polynomials, 168
   simplifying, 171-172
equilateral triangles, 230
equivalent fractions, 50-52
  reducing, 52-53
estimating angles, 219
evaluating expressions, 153-154
exponential notation, 96

exponents
  financial math, 131
    endowments, 132
    interest, 131
    investments, 131
    mortgages, 133-134
  integral, 93-94
    applications, 94-96
    division, 103
    exponential notation, 96
    hidden bases, 104-106
    multiplication, 98-101
    order of operations, 97-98
  negative, 101-103
  positive/negative, simplifying fractions, 177
  rational, 118
    multiple exponents, 121-122
    negative exponents, 120-121
    powers and roots, 119-120
    square and cube roots, 119
  relatives, 134
    areas of squares, 135
    prime factors, 136
    special properties of one and zero, 136-137
  rules, 107-124
  simplifying fractions, 122
    fraction form, 122-123
    positive/negative exponents, 123-124
expressions, 141-144
  addition, combining like terms, 147-149
  multiplication, 149
    Distributive Law, 150
    FOIL method, 150-153
    monomials, 149
  operations, 145-146
  plugging in, 153-154
  polynomials, 146-147
  simplifying fractions, 176
    common factors, 176
    positive/negative exponents, 177
  translation, 154-155
  variables, names, 142-144

**F**

factor trees, 7
factoring numbers, 5-7
  greatest common factors, 9-10
  least common multiples, 10-12
  prime factorizations, 7-9
factors, 5
Fahrenheit, converting to Celsius, 155
finances, percents, 71
  commissions, 74-75
  interest, 71-72
  profits, 75
  sales tax, 73
  tips, 73-74
financial math, 131
  endowments, 132
  interest, 131
  investments, 131
  mortgages, 133-134
First, Outer, Inner, Last. *See* FOIL method
FOIL (First, Outer, Inner, Last) method, 150-153
fractions, 23, 48
  addition/subtraction, 58
    like denominators, 58-59
    unlike denominators, 59-61
  applications, 24
  complex, 57
  cross-multiplication, 178-180
  decimals and, 53-54
    repeating decimals, 55-56
    terminating decimals, 54-55
  equivalent, 50-52
    reducing fractions, 52-53
  improper, 48-50
  multiplication/division, 56-58
  number lines, 27-28
  percents and, 66-68
  proportions. *See* proportions
  simplifying, 176
    common factors, 176
    positive/negative exponents, 177

simplifying with exponents, 122
 fraction form, 122-123
 positive/negative exponents, 123-124
frequency tables, 285-287

# G-H

generalizing solutions, roots, 114-116
geometry, 191
 angles, 211
 90°, 219
 acute, 220-221
 complementary, 223
 concave, 226
 convex, 225
 estimating, 219
 measuring, 212-215, 219
 obtuse, 221
 paired, 222-224
 right, 222
 straight, 227
 calculating area, 248
 circles, 254-257
 parallelograms, 250-252
 rectangles, 248-249
 trapezoids, 252-253
 triangles, 254
 graphing equations, 197-200
 measuring distance, 201
 parallel lines, 227-228
 perpendicular lines, 227-228
 plotting points, 192
 Cartesian plane coordinates, 192-197
 proportions, 265
 congruent shapes, 266
 scale factors, 268-276
 similar shapes, 267-269
 Pythagorean theorem, 244-245
 shapes, 229
 perimeter, 242-243
 polygons, 237-242
 quadrilaterals, 233-237
 triangles, 230-233

solids
 surface area, 257-264
 volume, 257-264
symmetry, 203
 glide reflection, 209
 mirror, 207-209
 rotation, 206-207
 translation, 204-206
glide reflection, 209
graphs
 equations, 197-200
 plotting points with Cartesian plane
 coordinates, 192-197
greatest common factors, 9-10
grouping symbols, order of operations, 32

hidden bases, integral exponents, 104-106
histograms, 292
hypotenuse, 232

# I-J

identity elements, 41-42
imperfect squares, calculating square roots,
 111-114
improper fractions, 48-50
independent events, probability, 304
 cards, 310-311
 coins, 304-307
 dice, 307-309
integers, 12-13
 signs, 14
 addition/subtraction, 14-16
 multiplication/division, 16
integral exponents, 93
 applications, 94
 exponential notation, 96
 repeated multiplication, 94-96
 division
 hidden bases, 104-106
 same bases, 103

multiplication
  exceptions, 101
  negative bases, 99-101
  positive bases, 98-99
  same bases, 98
  order of operations, 97-98
interest
  financial math, 131
  percents, 71-72
interior angles, polygons, 239-242
inverse operations, 43, 160
investments, financial math, 131
irrational numbers, 28-30
irregular polygons, 239
isosceles triangles, 230

## K–L

kite, 236

laws of exponents, 107-124
laws of operations
  Associative Law, 36-38
  Commutative Law, 34-36
  Distributive Law, 38
    minus sign, 39-41
    standard problems, 38
leading coefficients, 147
least common multiples, 10-12
length, scale factors, 271-276
like denominators, adding/subtracting fractions, 58-59
like terms, combining with equations, 164-165
line plots, 297
lowest common denominators, 59
lowest terms, reducing fractions, 52

## M

markup, 75
mathematical language (expressions), 141-144
  addition, combining like terms, 147-149
  multiplication, 149
    Distributive Law, 150
    FOIL method, 150-153
    monomials, 149
  operations, 145-146
  plugging in, 153-154
  polynomials, 146-147
  simplifying fractions, 176
    common factors, 176
    positive/negative exponents, 177
  translation, 154-155
  variables, names, 142-144
mean, 298-300
measurements
  angles, 212-215
    protractors, 215-219
  central tendency, 298
    mean, 298-300
    median, 300-301
    mode, 301
  distance, 201
  errors, 76
median, 300-301
minus sign, Distributive Law, 39-41
mirror symmetry, 207-209
mode, 301
monomials, 146, 149
more than one operation equations, 163
mortgages, financial math, 133-134
multi-step equations, 163
  combining like terms, 164-165
  more than one operation, 163
multiple exponents, 121-122
multiple variables, 173-174
multiples, 5

multiplication
  cross-multiplication, 51
  equations, 162-163
  expressions, 149
    Distributive Law, 150
    FOIL method, 150-153
    monomials, 149
  fractions, 56-58
  integers, 16
  integral exponents
    exceptions, 101
    negative bases, 99-101
    positive bases, 98-99
    same bases, 98
  order of operations, 33-34
multiplicative relationships, proportions, 79-81
  applications, 82-84
  science applications, 89
  unit multipliers, 84-87

**N**

names, variables, 142-144
natural numbers, 4-5
negative bases, 99-101
negative exponents, 101-103
  rational exponents, 120-121
  simplifying fractions, 123-124, 177
notations
  absolute value, 13
  cube roots, 115
  exponential, 96
  pi, 255
  radicals, 113
  square root, 109
number lines, 25
  absolute value, 13-14
  decimals, 26-27
  fractions, 27-28
  numerical comparisons, 25-26

numbers, 3
  base-ten number system, 126-127
  factoring, 5-7
    greatest common factors, 9-10
    least common multiples, 10-12
    prime factorizations, 7-9
  integers, 12-13
    signs, 14-16
  irrational, 28-30
  natural, 4-5
  number lines, absolute value, 13-14
  prime, 6
  rational, 22, 28-30
    repeating decimals, 22
    terminating decimals, 22
  real, 21
  scientific notation, 127
    large numbers, 128
    significant figures, 128
    small numbers, 129-130
  whole, 146
numerator (fractions), 24
numerical comparisons, 25-26

**O**

obtuse angles, 221
one operation equations, 160-161
operations
  expressions, 145-146
  inverse operations, 43, 160
  laws of
    Associative Law, 36-38
    Commutative Law, 34-36
    Distributive Law, 38-41
  one operation equations, 160-161
  order of, 32
    grouping symbols, 32
    integral exponents, 97-98
    multiplications/divisions, 33-34
    subtraction, 15

order of operations, 32
  grouping symbols, 32
  integral exponents, 97-98
  multiplications/divisions, 33-34
  subtraction, 15
ordinates, 193
origin, 193
outliers, 289-290

# P

paired angles, 222
  complementary, 223
  supplementary, 224
parallel lines, 227-228
parallelograms, 234
  calculating area, 250-252
pattern, decimals, 20
PEDMAS acronym (order of operations), 97
percents, 63
  calculations, 68
    finding percentages, 68-70
    percent change, 70-71
  decimals and, 64-66
  defined, 64
  financial examples, 71
    commissions, 74-75
    interest, 71-72
    profits, 75
    sales tax, 73
    tips, 73-74
  fractions and, 66-68
  percent change, 70-71
  percent error, 76-77
perfect squares, 110-111
perimeter (shapes), 242-243
perpendicular lines, 227-228, 250
pi, area of circles, 255-256
pictographs, 290
Pisano, Leonardo, 126

planes
  distance formula, 201
  graphing equations, 197-200
  plotting points with Cartesian plane
    coordinates, 192-197
  symmetry, 203
    glide reflection, 209
    mirror, 207-209
    rotation, 206-207
    translation, 204-206
plotting points with Cartesian plane
  coordinates, 192-197
polygons, 237
  interior angles, 239-242
  irregular, 239
  regular, 238
polynomials
  expressions, 146-147
  solving equations, 168
positive bases, multiplying exponents, 98-99
positive exponents, simplifying fractions,
  123-124, 177
powers, rational exponents, 119-120
prime factors, 136
prime factorizations, 7-9
prime numbers, 6
principal, calculating interest, 71
prisms, volume and surface area, 259-261
probability, 303
  chances of event not happening, 311
  independent events, 304
    cards, 310-311
    coins, 304-307
    dice, 307-309
profit margin, 75
profits, percents, 75
properties, variables, 144
proportions, 79-81, 265
  applications, 82
    recipes, 82-84
  congruent shapes, 266

scale factors, 268-269
  applications, 270
  area, 276
  length, 271-276
science applications, density, 89
similar shapes, 267-269
unit multipliers, 84
  currency conversions, 85-86
  unit conversions, 87
protractors, 215, 219
pyramids, volume and surface area, 261-262
Pythagorean theorem, 244-245

## Q

quadrilaterals, 233-237
qualitative data, 284-285
quantifying probability, 305
quantitative data, 284-285
quantitative relationships, expressions, 141-144
  addition, 147-149
  multiplication, 149-153
  operations, 145-146
  plugging in, 153-154
  polynomials, 146-147
  translation, 154-155
  variables, 142-144

## R

radians, measuring angles, 215
radicals, calculating square roots, 114
radius (circles), 256
rate-of-work problems, 180-182
rational exponents, 118-119
  multiple exponents, 121-122
  negative rational exponents, 120-121
  powers and roots, 119-120
  square and cube roots, 119

rational numbers, 22, 28-30
  repeating decimals, 22
  terminating decimals, 22
rationalizing the denominator, square roots, 116
  single square roots, 116-117
  two square roots, 117-118
ratios, 79-81
  recipe applications, 82-84
  science applications, density, 89
  unit multipliers, 84
    currency conversions, 85-86
    unit conversions, 87
real numbers, 21
recipes, proportions, 82-84
reciprocals, 57
recording data, frequency tables, 285-287
rectangles, 233
  calculating area, 248-249
reducing fractions, 52-53
regular polygons, 238
relationships
  additive, 79
  multiplicative
    applications, 82-84
    science applications, 89
    unit multipliers, 84-87
  proportions, 79-81
    applications, 82-84
    science applications, 89
    unit multipliers, 84-87
  quantitative expressions, 141-155
relatives, 134
  areas of squares, 135
  prime factors, 136
  special properties of one and zero, 136-137
remainder (factoring numbers), 5
repeated multiplication, integral exponents, 94-96
repeating decimals, 22, 55-56

resizing, scale factors, 268-269
  applications, 270
  area, 276
  length, 271-276
rhombus, 235
right angles, 222
right triangles, 232
roots
  generalizing the solution, 114-116
  rational exponents, 119-120
  square roots, 108-109
    calculating, 109-114
    rationalizing the denominator, 116-118
rotation symmetry, 206-207
rules of exponents, 107-124

## S

sales tax, percents, 73
scale factors, 84, 268-269
  applications, 270
  area, 276
  length, 271-276
scalene triangles, 231
science applications for proportions, density, 89
scientific notation, 127
  large numbers, 128
  significant figures, 128
  small numbers, 129-130
shapes (geometry), 229
  perimeter, 242-243
  polygons, 237
    interior angles, 239-242
    irregular, 239
    regular, 238
  quadrilaterals, 233-237
  triangles, 230-233
significant digits, 128
significant figures, scientific notation, 128
signs, integers, 14
  addition/subtraction, 14-16
  multiplication/division, 16

similar shapes, 267-269
simplest terms, reducing fractions to, 52
simplifying equations, 171-172
simplifying fractions, 176
  common factors, 176
  exponents, 122-124
  positive/negative exponents, 177
single square roots, rationalizing the denominator, 116-117
single variables, solving equations, 172-173
skip-counting, 5
solids
  surface area, 257
    cones, 263-264
    cubes, 258
    cylinders, 262-263
    prisms, 259-261
    pyramids, 261-262
  volume, 257
    cones, 263-264
    cubes, 258
    cylinders, 262-263
    prisms, 259-261
    pyramids, 261-262
solutions (equations), 159
  checking work, 172-174
  more than one solution, 167-168
  multiple variables with various powers, 169-171
  polynomials, 168
  simplifying, 171-172
sorting data, 287
  bins, 288-289
  outliers, 289-290
square roots, 108-109
  calculating, 109
    imperfect squares, 111-114
    perfect squares, 110-111
    radicals, 114
  rational exponents, 119
  rationalizing the denominator
    single square roots, 116-117
    two square roots, 117-118

squares, 135, 234
squaring a number, 135
standard notation, 128-129
standard problems, Distributive Law, 38
statistics, 303
stem-and-leaf plots, 294-295
straight angles, 227
subtraction
    equations, 161-162
    fractions, 58
        like denominators, 58-59
        unlike denominators, 59-61
    integers, 14-16
    order of operations, 15
supplementary angles, 224
surface area of solids, 257
    cones, 263-264
    cubes, 258
    cylinders, 262-263
    prisms, 259-261
    pyramids, 261-262
symbols, grouping (order of operations), 32
symmetry, 203
    glide reflection, 209
    mirror, 207-209
    rotation, 206-207
    translation, 204-206

**T**

tax rate, 73
temperature conversions, 155
terminating decimals, 22
    fractions and, 54-55
terms, 146
theorems, Pythagorean theorem, 244-245
tips, percents, 73-74
translation
    expressions, 154-155
    symmetry, 204-206
trapezium, 236
trapezoids, 235, 252-253
triangles, 230-233, 254
two square roots, rationalizing the denominator, 117-118

**U**

unit conversions, 87
unit multipliers, proportions, 84
    currency conversions, 85-86
    unit conversions, 87
units, 81
unlike denominators, adding/subtracting fractions, 59-61

**V**

variables, 142
    equations
        more than one variable, 165-167
        multiple variables with various powers, 169-171
    names, 142-144
Venn diagrams, 295-297
volume, solids, 257
    cones, 263-264
    cubes, 258
    cylinders, 262-263
    prisms, 259-261
    pyramids, 261-262

**W-X-Y-Z**

whole numbers, 146
word problems, 175
    distance-rate-time problems, 182-188
    rate-of-work problems, 180-182

x-axis, 193
x-coordinate, 195

y-axis, 193
y-coordinate, 195